Studies in Fuzziness and Soft Computing

Volume 362

Series editor

Janusz Kacprzyk, Polish Academy of Sciences, Warsaw, Poland
e-mail: kacprzyk@ibspan.waw.pl

The series "Studies in Fuzziness and Soft Computing" contains publications on various topics in the area of soft computing, which include fuzzy sets, rough sets, neural networks, evolutionary computation, probabilistic and evidential reasoning, multi-valued logic, and related fields. The publications within "Studies in Fuzziness and Soft Computing" are primarily monographs and edited volumes. They cover significant recent developments in the field, both of a foundational and applicable character. An important feature of the series is its short publication time and world-wide distribution. This permits a rapid and broad dissemination of research results.

More information about this series at http://www.springer.com/series/2941

Robert John · Hani Hagras
Oscar Castillo

Editors

Type-2 Fuzzy Logic
and Systems

Dedicated to Professor Jerry Mendel for his
Pioneering Contribution

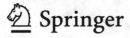 Springer

Editors
Robert John
LUCID Research Group, School
    of Computer Science
The University of Nottingham
    Jubilee Campus
Nottingham, Nottinghamshire
UK

Oscar Castillo
Division of Graduate Studies
Tijuana Institute of Technology
Tijuana, Baja California
Mexico

Hani Hagras
School of Computer Science
    and Electronic Engineering
The Computational Intelligence Centre
University of Essex
Colchester
UK

ISSN 1434-9922   ISSN 1860-0808   (electronic)
Studies in Fuzziness and Soft Computing
ISBN 978-3-319-89218-4   ISBN 978-3-319-72892-6   (eBook)
https://doi.org/10.1007/978-3-319-72892-6

Printed on acid-free paper

This Springer imprint is published by Springer Nature
The registered company is Springer International Publishing AG
The registered company address is: Gewerbestrasse 11, 6330 Cham, Switzerland

Preface

This book is dedicated to Jerry, Prof. Jerry Mendel, for his pioneering works on the type-2 fuzzy sets and systems. Jerry has had a long and distinguished academic career in both signal processing and fuzzy logic, winning many awards and honours. However, since his first paper in 1998 he has been the leader in the fascinating area of the type-2 fuzzy logic field for nearly 20 years. He has over 40 journal articles in leading journals with many citations. His three most cited papers in Google Scholar (as for July 2017) are: "Type-2 fuzzy sets made simple", J. M. Mendel, R. I. John, IEEE Transactions on Fuzzy Systems 10 (2), 117–127, 2002 (1788 citations), "Interval type-2 fuzzy logic systems: theory and design", Q. Liang, J. M. Mendel, IEEE Transactions on Fuzzy Systems 8 (5), 535–555, 2000 (1380 citations) and "Type-2 fuzzy logic systems", N. N. Karnik, J. M. Mendel, Q. Liang, IEEE transactions on Fuzzy Systems 7 (6), 643–658, 1999 (1200 citations).

Jerry has worked with numerous Ph.D. students and colleagues from across the world, always in a collaborative way to move the field forward. We have had many long hours discussing important research issues in type-2 fuzzy logic. Over that time he has become our friend and we are honoured to put together this invited collection of contributions.

The chapters here cover a wide variety of topics—the type-2 fuzzy sets and the game Go, weighted averages, control of agricultural vehicles, challenges for the type-2 fuzzy control, type-2 fuzzy control in games, pattern recognition and the role of type-2 fuzzy sets in intelligent agents, just to mention a few.

We would like to thank the authors for their interesting contributions. The diversity of topics covered and views and perspectives presented reflects the diversity in the type-2 community. If you are new to type-2 fuzzy logic, we hope you are inspired to read these and follow up on Jerry's work.

Nottingham, UK   Robert John
Colchester, UK   Hani Hagras
Tijuana, Mexico   Oscar Castillo
Spring 2017

Contents

From T2 FS-Based MoGoTW System to DyNaDF for Human and Machine Co-learning on Go

Chang-Shing Lee, Mei-Hui Wang, Sheng-Chi Yang
and Chia-Hsiu Kao

Abstract This chapter describes the research from T2 FS-based MoGoTW system to DyNamic DarkForest (DyNaDF) open platform for human and machine co-learning on Go. A human Go player's performance could be influenced by some factors, such as the on-the-spot environment as well as physical and mental situations of the day. In the first part, we used a sample of games played against machine to estimate the human's strength (Lee et al. in IEEE Trans Fuzzy Syst 23 (2):400–420, 2015 [1]). The Type-2 Fuzzy Sets (T2 FSs) with parameters optimized by a genetic algorithm for estimating the rank was presented (Lee et al. in IEEE Trans Fuzzy Syst 23(2):400–420, 2015 [1]). The T2 FS-based adaptive linguistic assessment system inferred the human performance and presented the results using the linguistic description (Lee et al. in IEEE Trans Fuzzy Syst 23(2):400–420, 2015 [1]). In March 2016, Google DeepMind challenge match between AlphaGo and Lee Sedol in Korea was a historic achievement for computer Go development. In Jan. 2017, an advanced version of AlphaGo, Master, won 60 games against some top professional Go players. In May 2017, AlphaGo defeated Ke Jie, the top professional Go player, at the Future of Go Summit in China. In second part, we showed the development of computational intelligence (CI) and its relative strength in comparison with human intelligence for the game of Go (Lee et al. in IEEE Comput Intell Mag 11(3):67–72, 2016 [2]). Additionally, we also presented a robotic prediction agent to infer the winning possibility based on the information generated by DarkForest Go engine and to compute the winning possibility based on the partial game situation inferred by FML assessment engine (Lee et al. in FML-based prediction agent and its application to game of Go, 2017 [3]). Moreover, we chose seven games from 60 games to evaluate the performance (Lee et al. in FML-based prediction agent and its application to game of Go, 2017 [3]). In this chapter, we extract the human domain knowledge from Master's 60 games for giving the desired output. Then, we combine Particle Swarm Optimization (PSO) and FML to learn the knowledge base and further infer the game results of Google AlphaGo in May 2017. The experimental results show that the proposed approach is feasible for

C.-S. Lee (✉) · M.-H. Wang · S.-C. Yang · C.-H. Kao
National University of Tainan, Tainan, Taiwan
e-mail: leecs@mail.nutn.edu.tw

© Springer International Publishing AG 2018
R. John et al. (eds.), *Type-2 Fuzzy Logic and Systems*,
Studies in Fuzziness and Soft Computing 362,
https://doi.org/10.1007/978-3-319-72892-6_1

the application to human and machine co-learning on Go. In the future, powerful computer Go programs such as AlphaGo are expected to be instrumental in promoting Go education and AI real-world applications.

1   Introduction

Many real-world applications are with a high-level of uncertainty. Type-2 FS (T2 FS) has the ability to capture the uncertainty about membership functions of fuzzy sets [4, 5]. Moreover, Type-2 Fuzzy Logic System (T2 FLS) is used to handle the high uncertainties in the group decision-making process as it can model the uncertainties between expert preferences by using T2 FSs [4–7]. Because of the popularity of T2 FS [7], the Type-2 Fuzzy Markup Language (T2 FML), an extension of the FML grammar, is developed to allow system designers to express their expertise by using an Interval Type-2 Fuzzy Logic System (IT2 FLS) to model type-2 fuzzy sets and systems [8–10]. Fuzzy Markup Language (FML) has become an IEEE Standard since May 2016 and provides designers of intelligent decision making systems with a unified and high-level methodology for describing systems' behaviors by means of rules based on human domain knowledge [8, 10]. FML is a fuzzy-based markup language that can manage fuzzy concepts, fuzzy rules, and a fuzzy inference engine [8, 10]. Additionally, FML is with the following features: understandability, extendibility, and compatibility of implemented programs as well as efficiency of programming [10]. The main advantage of using FML is easy to understand and extend the implemented programs for other researchers [8, 10].

The game of Go is played by two players, *Black* and *White*. Two Go players alternatively play their stone at a vacant intersection of the board by following the rules of Go [11]. Additionally, Go is regarded as one of the most complex board games because of its high state-space complexity 10^{171}, game-tree complexity 10^{360}, and branching factor 250 [12]. The skill of amateur players in Go is ranked according to *kyu* (K) in the lower tier, where a smaller number stands for stronger playing skill (with 1 K being the highest skill level), and *dan* (D) in the higher tier, where a larger number stands for stronger playing skill. Professional Go players are ranked entirely in dan, abbreviated with the letter P [2]. Go is typically played on 19×19 size boards, but 9×9 size boards are also common for beginners. The complexity of the 9×9 game is far less than the standard game, and the 9×9 game had been one of the interim goals for computer Go programs [2]. The handicaps for the human vs. computer 19×19 game have been decreased from 29 in 1998 to 0 in 2016 [2]. In May 2017, AlphaGo even defeated Ke Jie, the top professional Go players in the world, at the Future of Go Summit in Wuzhen [13]. Owning to the quick advance in artificial intelligence, currently powerful computer Go programs such as AlphaGo [14] and DeepZenGo are expected to give top professional humans a few handicap stones to make for an even match.

Games have served as one of the best benchmarks for studying artificial intelligence [15, 16]. Over the last few years, Monte Carlo Tree Search (MCTS) has already made a profound effect on artificial intelligence, especially in computer games [15]. Gelly and Silver [17] applied Rapid Action Value Estimation (RAVE) algorithm and Heuristic MCTS to a computer Go program, MoGo. Monte Carlo tree search (MCTS), minorization-maximization (MM), and deep convolutional neural networks (DCNNs) have demonstrated great success in Go [2, 14, 18, 19]. In December of 2014, two teams applied deep convolutional neural networks to Go independently [20, 21]. Among many of DCNN's applications, it has seen success in image and video recognition. When applied to Go, DCNN is able to recognize move patterns at a significantly lower error rate than MM. For this reason, most state-of-the-art computer Go programs use MCTS combined with either MM or DCNN [2].

For evaluating the human performance on Go games, humans could be advanced to a higher rank based on the number of winning games via a formal human against human competition [1]. However, the invited human Go player's strength might be affected by some factors, such as the on-the-spot environment, physical and mental situations of the day, and game settings, so the Go player's rank may be with an uncertain possibility. Additionally, one player's strength may gradually decrease because of getting older or seldom playing with a stronger human [1]. Hence, these uncertain factors cause the difficulties and uncertainty in evaluating the rank of one human Go player. In [1], we used T2 FSs to model the requirements of a person specification that is reflective of all the experts' opinions and this can be used to provide a good evaluation for the rank of the Go players. A T2 FS-based adaptive linguistic assessment system was proposed to evaluate one human Go player's performance with a semantic analysis such that the proposed system is helpful to increase the human Go player's enthusiasm for playing with the computer Go program [1], especially for children.

In [2], we helped the readership better understand how the development of computer Go programs has arrived at this milestone of winning against one of the top human players, and how IEEE Computational Intelligence Society (CIS) has been involved in this process. This huge achievement in AI is based largely on CI methods, including DCNNs, supervised learning from expert games, reinforcement learning, the use of the value network and policy network, and MCTS. In [3], we constructed a DyNamic DarkForest (DyNaDF) Cloud Platform for game of Go, including a demonstration game platform, a machine recommendation platform, and an FML assessment engine. We used the first-stage prediction results of DarkForest Go engine [18, 19] and the second-stage inferred results of the FML assessment engine [22], we further introduced the third-stage FML-based decision support engine to predict the winner of the game and chose seven games from Master's 60 games in Jan. 2017 [3, 23] to evaluate the performance. The fourth-stage robot engine reports real-time situation to players. This chapter further combines FML and particle swarm optimization (PSO), called PFML [24], to learn the domain knowledge of Master's 60 games [23] by referring to the book published in Taiwan [25]. After learning, we use the proposed approaches in [3] to infer the game results of the Future of Go Summit in Wuzhen in May 2017. From the experimental results, we can get much higher accuracy than before learning.

The remainder of this chapter is organized as follows: Sect. 2 introduces the research performance from the proposed T2 FS-based MoGoTW system [1, 2] to the FML-based DyNaDF open system [3, 22]. Section 3 is dedicated to the human and machine co-learning part based on T2 FS and FML. Section 4 shows some experimental results. Finally, conclusions are given in Sect. 5.

2 From T2 FS-Based MoGoTW Linguistic Assessment System to FML-Based DyNamic DarkForest Open Platform

This section introduces the research performance from the proposed T2 FS-based MoGoTW linguistic assessment system [1] to the constructed FML-based DyNaDF open platform [3].

2.1 T2 FS-Based MoGoTW System for Adaptive Linguistic Assessment

Over the past years, there were many Go competitions between humans and computer Go programs held in Taiwan or in the world [2, 26]. However, playing with the computer Go program may be boring because the computer Go program cannot express its feelings, especially in one lopsided game [27, 28]. If the computer Go program is able to adaptively assess its opponent's strength and provide one real-time feedback mechanism for Go players, it will be helpful for humans to increase their interest in playing with the computer Go program and to find their relevant opponents and/or relevant handicap. Upper Confidence Bounds for Trees (UCT) is the most popular algorithm in the MCTS family [15, 29, 30]. MoGoTW, developed based on MoGo 4.86 Sessions plus the Taiwan (TW) modifications, plays its move at the board according to the result of the best-move selection mechanism [1]. The strength of MoGoTW is increased when it loses and is decreased when it wins based on item response theory (IRT) [31].

In [1], we proposed the T2 FS-based adaptive linguistic assessment system to evaluate human Go player's performance whose structure is shown in Fig. 1. The MCTS *Simulation Number (SN)* is adjusted to meet the strength of the opponent during one round and their operations are described as follows: (1) **Adjustment in per-move** SN: SN is increased by multiplying by V_1 when the *Winning Rate (WR)* of the computer program is less than WR_1. (2) **Adjustment in per-game** SN: When one game ends and the computer program wins the game, the computer program weakens its strength by dividing SN into V_2 for next game. (3) The involved human Go players compete K games against MoGoTW for one round. During the competition, MoGoTW adjusts its strength to match with the strength of the human Go

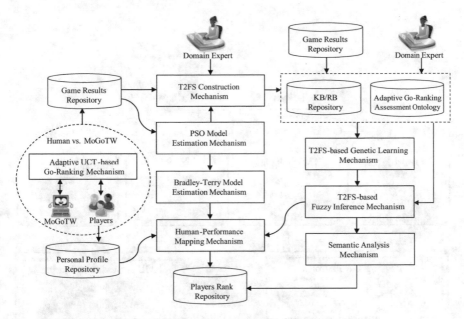

Fig. 1 T2 FS-based adaptive linguistic assessment system [1]

player by increasing or decreasing MCTS's simulation number after playing one move and one game. (4) If human wins the first game, then MoGoTW increases MCTS's simulation number to strengthen its own strength at the start of the second game to compete with the human. In other words, the more consecutive games are won by the human, the stronger the human. *GameWeight* denotes the strength of all the games played by this human. The higher *GameWeight*, the stronger the human. *WinningRate* denotes the winning rate of the human after playing games with MoGoTW. Based on this concept, T2 FSs $\widetilde{GameWeight}_{Low}$, $\widetilde{GameWeight}_{Medium}$, $\widetilde{GameWeight}_{High}$ $\widetilde{WinningRate}_{Low}$, $\widetilde{WinningRate}_{Medium}$, and $\widetilde{WinningRate}_{High}$ are constructed.

2.2 FML-Based DyNamic DarkForest (DyNaDF) Open Platform

Figure 2 shows the structure of the DyNaDF open Platform whose brief descriptions are given as follows [3]: (1) The DyNaDF open platform for game of Go application is composed of a playing-Go platform located at National University of Tainan (NUTN)/Taiwan and National Center for High-Performance Computing (NCHC)/Taiwan, a DarkForest Go engine located at Osaka Prefecture University

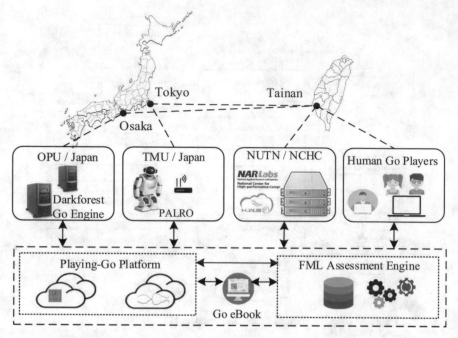

Fig. 2 Structure of DyNamic DarkForest open platform for Go [3]

(OPU)/Japan, and the robot PALRO from Tokyo Metropolitan University (TMU)/ Japan. (2) Human Go players surf on the DyNaDF platform located at NUTN/ NCHC to play with DarkForest Go engine located in OPU. (3) The FML assessment engine infers the current game situation based on the prediction information from DarkForest and stores the results into the database. (4) The robot PALRO receives the game situation via the Internet and reports to the human Go players. Human can learn more information about game's comments via Go eBook [3]. Figure 3 shows the screenshot of the game between Ke Jie (9P) as Black and Master as White on Dec. 30, 2016 provided by the FML-based DyNaDF Open Platform [3].

3 Human and Machine Co-learning Based on T2 FS and FML

This section introduces the human and machine co-learning based on T2 FS and FML. Section 3.1 describes the adaptive human performance evaluation on game of Go. The FML-based prediction agent for DyNaDF open platform is described in Sect. 3.2.

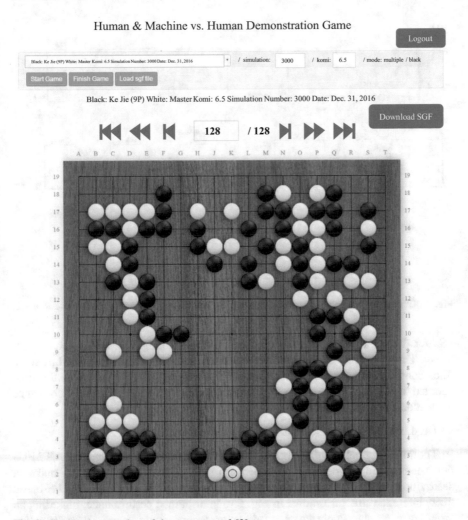

Human & Machine vs. Human Demonstration Game

Fig. 3 Captured screenshot of the game record [3]

3.1 Adaptive Human Performance Evaluation on Game of Go

- **T2 FS-Based FML for Knowledge Base Representation**

In [1], we presented the novel human performance knowledge representation for game of Go based on fuzzy sets and fuzzy markup language. The constructed fuzzy sets for the Go competition were stored in the adaptive Go-ranking assessment ontology. In addition, the refined concept for the fuzzy knowledge base was presented by a fuzzy linguistic label set and a fuzzy data set. Figures 4a–e show the T2

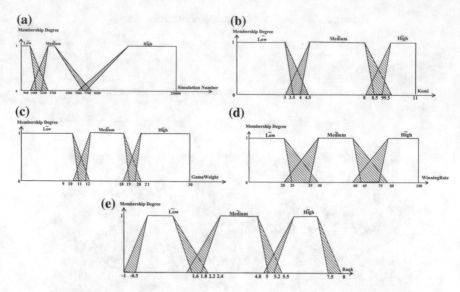

Fig. 4 T2 FSs for fuzzy variables **a** *SN*, **b** *Komi*, **c** *GameWeight*, **d** *WinningRate* and **e** *Rank* [1]

FSs for fuzzy variables *SN*, *Komi*, *GameWeight*, *WinningRate*, and *Rank*, respectively. There are total 81 fuzzy inference rules in the rule base of FML, but some rules conflict with the fact. After removing these conflicted fuzzy rules, the final adopted fuzzy rules are 63 in the application to adaptive human performance evaluation on game of Go [1].

- **Six-Layer T2 FS-Based Fuzzy Inference Structure**

The difference between T1 FLS and T2 FLS is the output processing [6]. A T2 FLS first needs to do a type-reducer to perform the type-reduction and then makes a defuzzification on the type-reduced set [4, 6]. We proposed a six-layer T2 FS-based fuzzy inference structure shown in Fig. 5 to infer human performance, including an input layer, an antecedent layer, a rule layer, a consequent layer, a type-reduction layer, and an output layer [1]. The descriptions for each layer are as follows: (1) The **input layer** represents the values of the input variables, such as *GameWeight* (*GW*), *WinningRate* (*WR*), *SN*, and *Komi* for one round and they are directly transmitted to the antecedent layer, for example, $x' = \{GW', WR', SN', Komi'\}$. (2) The **antecedent layer** is to compute the membership of the input values from the first layer. However, for an IT2 FS, the membership is an interval, for example, the membership of $\tilde{SN}_{Low}$ is $[\mu_{\underline{SN}_{Low}}(SN), \mu_{\overline{SN}_{Low}}(SN)]$. (3) In the **rule layer**, fuzzy matching uses fuzzy AND operator (MIN) in the antecedent to calculate the degree to which the input data match the condition of the *M* fuzzy rules. (4) In the **consequent layer**, two iterative Karnik-Mendel (KM) algorithms [4, 6] are used to compute the left-end and right-end points of the centroid of each rue's consequent IT2 FS. (5) In the **type-reduction layer**, center-of-sets type reduction combines an individual fired interval with its

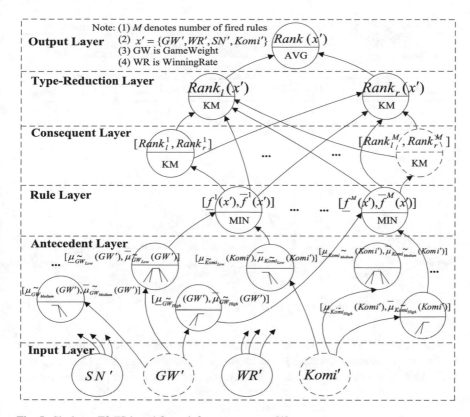

Fig. 5 Six-layer T2 FS-based fuzzy inference structure [1]

corresponding pre-computed consequent left-end and right-end centroids, and then implements two iterative KM algorithms to perform the type reduction set to generate the lower point and upper point. (6) In the **output layer**, the crisp output of the IT2 FS, $Rank(x')$, is defuzzied by averaging $Rank_l(x')$ and $Rank_r(x')$.

- **Machine Learning Mechanism for Adaptive Assessment System**

The genetic learning mechanism is adopted to tune the fuzzy linguistic label of T2 FS [1]. The brief descriptions of the machine learning mechanism in [1] are as follows: (1) The genes of the encoded chromosome contain three parts, including the *knowledge-based genes*, the *rule-based genes*, and the *linguistic-hedged genes*. The *knowledge-based genes* represent the linguistic labels with data and parameters of fuzzy variables, *rule-based genes* store the domain expert's weights and updated weights of the fuzzy inference rules, and *linguistic-hedged genes* denote the fuzzy linguistic hedge such as *normal*, *more-or-less*, or *very*. (2) One chromosome is composed of 73 genes, including G_1–G_5 are the *knowledge-based genes*, G_6–G_{68} are the *rule-based genes*, and G_{69}–G_{73} are the *linguistic-hedge genes*. *Mean Square Error* (*MSE*) is the fitness function. Figure 6 shows one chromosome with three parts of the

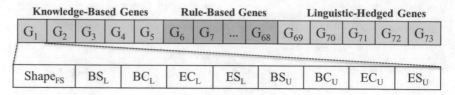

Fig. 6 One chromosome with three parts of the genes [1]

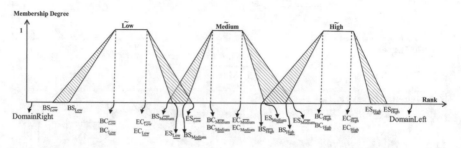

Fig. 7 T2 FS Restrictions for the genetic tuning [1]

genes. Figure 7 shows the graphic representation for the genetic tuning the T2 FS *Rank*. The restrictions on tuning the linguistic labels *Medium* and *High* of fuzzy variable *Rank* are similar to tuning the linguistic label *Low* of fuzzy variable *Rank*.

3.2 FML-Based Prediction Agent for DyNaDF Open Platform

Figure 8 shows the proposed FML-based prediction agent for DyNaDF open platform, including *Stage I: DarkForest Go engine*, *Stage II: FML assessment engine*, *Stage III: PFML learning mechanism* [24], *Stage IV: FML-based decision*

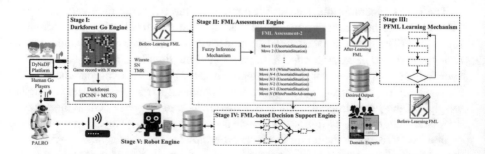

Fig. 8 System structure of five-stage FML-based prediction agent [3]

support engine, and *Stage V: robot engine* [3]. The followings are its short descriptions:

- **Stage I**: We applied Facebook DarkForest Go open source that trains a DCNN to predict the next top-k moves [18]. DarkForest Go engine powered by deep learning has been developed mainly by Tian and Zhu from Facebook AI Research (FAIR) since May 2015 and was open to the public in 2016 [18, 19]. DarkForest relies on a DCNN designed for long-term predictions and has been able to substantially improve the winning rate for pattern matching approaches against MCTS-based approaches, even with looser search budgets [18]. Tian and Zhu [19] proposed a 12-layered full convolutional network architecture for DarkForest where (1) each convolution layer is followed by a ReLU nonlinearity, (2) all layers use the same number of filters at convolutional layers ($w = 254$) except for the first layer, (3) no weight sharing is used, (4) pooling is not used owing to negatively affecting the performance, and (5) only one softmax layer is used to predict the next move of Back and White to reduce the number of parameters.

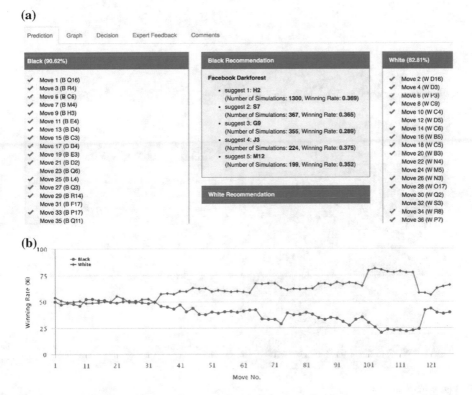

Fig. 9 Captured screenshots including **a** predicted next moves, **b** winning rate curve, **c** inferred game results, and **d** feedback from domain experts

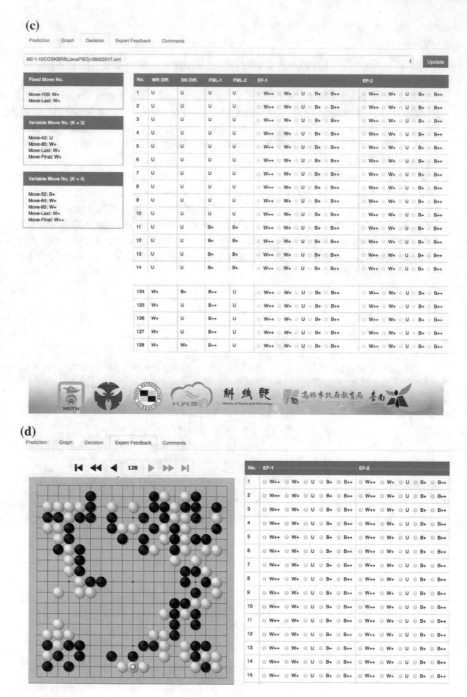

Fig. 9 (continued)

- **Stage II**: FML assessment engine adopted each-move-position, DarkForest-predicted top-5-move number of simulations and winning rate to decide each-move number of simulations (*BSN* and *WSN*), winning rate (*BWR* and *WWR*), and suggestion affect (*BSA* and *WSA*) [22]. After that, the FML assessment engine inferred each-move current game situation (*CGS*), including "*Black is obvious advantage (BlackObviousAdvantage, B^{++}),*" "*Black is possible advantage (BlackPossibleAdvantage, B^{+}),*" "*Both are in an uncertain situation (UncertainSituation, U),*" "*White is possible advantage (WhitePossibleAdvantage, W^{+}),*" and "*White is obvious advantage (WhiteObviousAdvantage, W^{++}).*" We use the game between Ke Jie as Black and Master as White on Dec. 30, 2016 as an example to show the captured screenshots, including the predicted next moves, winning rate curve, inferred game results, and feedback provided by the domain experts in Figs. 9a–d, respectively.
- **Stage III**: We extracted the domain knowledge from the book [25] to give the desired output to 60 Master's games in Jan. 2017 [23]. Table 1 shows the methods that give the desired output to a specific move and the other moves in the neighborhood of this specific move. In Stage III, the proposed PFML learning mechanism in [24] which combines PSO with FML, optimizes the knowledge base of the constructed FML with 5-fold cross validation. In this chapter, the number of generations is 3000 and the number of particles is 20.
- **Stage IV**: The proposed FML-based decision support engine computed the winning possibility based on the partial game situation inferred by FML assessment engine and stored the predicted results into the database. In this chapter, we use the proposed methods in [3], including sampling the information from three or four sub-games, to infer the final game result. Figure 10 gives an example of a game with 178 moves when sampling the information from four sub-games, where the input information of *Neighborhood$_{1/4}$*, *Neighborhood$_{2/4}$*, *Neighborhood$_{3/4}$*, and *Neighborhood$_{4/4}$* is the current game situations for *Neighborhood* of fuzzy number **Move-45** (moves 40–50), *Neighborhood* of fuzzy number **Move-90** (moves 85–95), *Neighborhood* of fuzzy number **Move-135** (moves 130–140), and *Neighborhood* of fuzzy number **Move-Last** (moves 168–178), respectively. For example, if the input vector is $x = (x_1, x_2, ..., x_{11})$, then vector $x_{i/k}$ denotes the input 11 current game situations (*CGSs*) in the *Neighborhood$_{i/k}$*. Figure 11 is the fuzzy sets for B^{++}, B^{+}, U, W^{+}, and W^{++} [3].
- **Stage V**: The robot engine retrieved information from the database to comment on the game situation, including (1) Black and White's move numbers that appear the first 3 highest and the last 3 lowest number of simulations as well as the highest and lowest winning rates, and (2) Black and White's average winning rates and top-move rates. It also reports the real-time predicted top-3-move positions to the human Go player to think carefully before playing his/his next move [22]. Figure 12 shows the diagram about how to program code to the robot PALRO, developed by Japan FUJISOFT Incorporated [3]. Table 2 shows the comment on the game between Ke Jie as Black and Master as White on Dec. 30, 2016.

Table 1 **a** Method giving the desired output to a specific move and **b** two methods giving the desired output to the neighborhood of a specific move

a

- If the domain experts [25] consider (1) Black move N to be a *Tesuji*, an *Excellent Move*, or a *Move of Master*, then the desired output of Black move N is given to B^{++}. On the contrary, the desired output of White move N is W^{++}, and (2) *White was lost* or *Black Establishes Competitive Advantage*, then the desired output of Black move N is given to B^{++}. On the contrary, the desired output of White move N is W^{++}
- If the domain experts consider (1) Black move N to be a *Surprising Move*, a *Sente*, a *Big Point*, a *Good Move*, or a *Severe Tactics*, then the desired output of Black move N is given to B^{+}. On the contrary, the desired output of White move N is W^{+}, and (2) Black move N to be a *Losing Move*, *Failure Move*, or a *Big Bad Move*, then the desired output of White move N was given to W^{+}. On the contrary, the desired output of Black move N is B^{+}

b

- **Method 1**: (1) Moves $N-5$, $N-4$, ...$N-1$, $N+1$, ..., to $N+5$: The same one as move N. (2) Last 11 moves of the game: The same one as the final exact game result
- **Method 2**: (1) Moves $N+1$, ..., to $N+5$: Each move increases by 0.1 for W^{+}/W^{++} and decreases by 0.1 for B^{+}/B^{++}. (2) Last 11 moves of the game: Each two moves increase by 0.1 for W^{+}/W^{++} and decrease by 0.1 for B^{+}/B^{++}. (3) First 100 moves: If domain experts consider move 1, 2,..., or 100 to be B^{++}/W^{++}, we only give B^{+}/W^{+} to its desired output

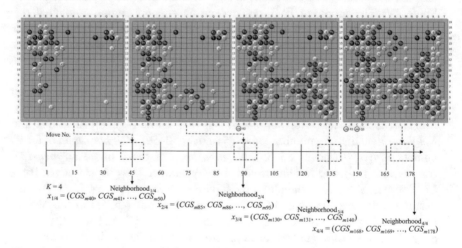

Fig. 10 Game with 178 moves when sampling the information from four sub-games [3]

4 Experimental Results

This section introduces the experimental results. We first describe our experimental results for MoGoTW in Sect. 4.1, including comparison between T1 FS and T2 FS and genetic learning performance based on T2 FS. Section 4.2 gives the experimental results of the constructed DyNaDF Open Platform with FAIR Open Source DarkForest.

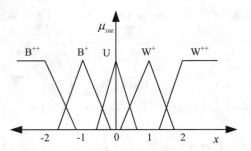

Fig. 11 Fuzzy sets for B^{++}, B^+, U, W^+, and W^{++} [3]

Fig. 12 Diagram about programming code to the robot PALRO

Table 2 Comments on the game [22]

- **Black**
  The first 3 highest simulation numbers occurred at Moves B17 (3149), B67(3097), and B43(3097). The last 3 lowest simulation numbers occurred at Moves B45(223), B77(231), and B101(235). The information of estimated possible winning rate: The highest winning rate is B13(52.38%), the lowest winning rate is B105(18.31%), and the average winning rate is 37.86%. Top-move Rate is 90.62%

- **White**
  The first 3 highest simulation numbers occurred at Moves W44 (3173), W68(3164), and W72(3157). The last 3 lowest simulation numbers occurred at Moves W88(204), W28(229), and W106(317). The information of estimated possible winning rate: The highest winning rate is W104(81.89%), the lowest winning rate is W12(48.37%), and the average winning rate is 62.25%. Top-move Rate is 82.81%

4.1 Experimental Results for MoGoTW

In order to evaluate the performance of the proposed approach in [1], we first invited the human Go players with Dan level or Kyu level to play with the adaptive-ability

Table 3 Before-learning *MSE*, 1-CS, and accuracy based on T1 FS and T2 FS [1]

Measures	T1 FS		T2 FS	
	Training data	Testing data	Training data	Testing data
MSE	0.44	0.34	0.33	0.26
1-*CS*	0.014	0.012	0.01	0.009
Accuracy (%)	86.6	94.3	91.6	94.3

MoGoTW for 9×9 games from 2012 to 2013 via the open held human against computer Go competitions at NUTN, IEEE WCCI 2012, and FUZZ-IEEE 2013 [2]. Second, all of the invited human Go players are *Black* and MoGoTW is *White*. Third, each human player played five consecutive games for each round, but each human player was encouraged to compete with MoGoTW as many rounds as possible. The MCTS simulation number is adjusted after each game or after each move [1].

- **Comparison Between T1 FS and T2 FS**

  The input data were divided into two parts, namely training data with 60 records and testing data with 53 records. Because human's certificated rank is an integer and not every player's performance is always expected as his/her certificated rank during the competition, the desired output is adjusted according to the real game situation, such as the searched maximum likelihood *SN*, *WinningRate*, and *GameWeight*. *MSE*, 1-*CS* (Cosine Similarity), and accuracy are the adopted measures in [1]. Accuracy is computed by dividing the number of the positive matched results over the number of total records. Table 3 indicates that T2 FS performs better than T1 FS and proves that T2 FS has an advantage to handle the high uncertainties in the world [1].

- **Genetic Learning Performance Based on T2 FS**

  Figure 13 shows the before-learning and after-learning *MSE* curves for T1 FS and T2 FS. N_{GEN} is 1000, 2000, 4000, or 8000. The pair of crossover rate and mutation rate is 0.75/0.05 and 0.65/0.1 for T1 FS and T2 FS, respectively. Observe Fig. 13 that: (1) Before learning, T2 FS performs better than T1 FS for both the training data and the testing data. (2) Before learning, the testing data performs better than training data, but after learning, *MSE* of training data is lower than *MSE* of testing data. This proves that the genetic learning is effective for the application in. (3) *MSE* decreases a lot after learning 1000 generations and the drop is much bigger for T1 FS than T2 FS. (4) *MSE* has a tendency to reduce when number of generations is increased, and *MSE* has an obvious drop when N_{GEN} is 2000. Figures 14a–e show the after-learning T2 FSs for fuzzy variables *SN*, *Komi*, *WinningRate*, *GameWeight*, and *Rank*, respectively, when crossover rate = 0.65 and mutation rate = 0.1.

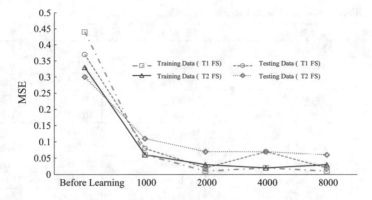

Fig. 13 Curves of MSE under different number of generations [1]

4.2 Experimental Results for DyNaDF Platform with FAIR Open Source DarkForest

In order to evaluate the performance of the constructed DyNaDF platform with FAIR open source DarkForest, we did the following steps: (1) the invited human Go players surfed on the DyNaDF cloud platform to play with DarkForest located in NUTN, NCHC, or OPU, (2) the game records on the Internet were downloaded and fed into the DyNaDF cloud platform, (3) the predicted each-move information was generated by DarkForest during playing, and (4) the each-move and final-move current game situation was inferred by the proposed approaches in [3].

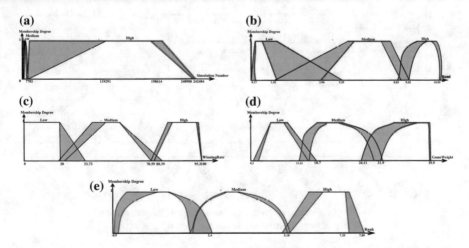

Fig. 14 After-learning T2 FSs for fuzzy variables **a** *SN*, **b** *Komi*, **c** *WinningRate*, **d** *GameWeight*, and **e** *Rank* under 0.65/0.1 [1]

Table 4 Basic profile of collected 10 games [3, 13, 23]

Game No.	Black/Level	White/Level	Date	Winner
G1	Master	Ke Jie/9P	2016/12/30	Black
G2	Ke Jie/9P	Master	2016/12/30	White
G3	Master	Yuta Iyama/9P	2017/1/2	Black
G4	Ke Jie/9P	Master	2017/1/3	White
G5	Chun-Hsun Chou/9P	Master	2017/1/4	White
G6	Master	Weiping Nie/9P	2017/1/4	Black
G7	Gu Li/9P	Master	2017/1/4	White
G8	Ke Jie/9P	AlphaGo	2017/5/23	White
G9	AlphaGo	Ke Jie/9P	2017/5/25	Black
G10	AlphaGo	Ke Jie/9P	2017/5/27	Black

- **FML-Based Prediction Agent**

Table 4 shows the basic profile of the collected 7 games between Master and top
professional Go players in Jan. 2017 (G1–G7) and 3 games between AlphaGo and
Ke Jie in May 2017 (G8–G10) [13, 23]. We fed the collected game records
downloaded from the Internet into our developed DyNaDF open platform. Figure 15 shows the DarkForest-predicted winning rate for G1–G7 on *Neighborhood*
of fuzzy number ***Move-100***, *Neighborhood* of fuzzy number ***Move-200***, and
Neighborhood of fuzzy number ***Move-Last***. Figure 15 also shows that DarkForest
successfully predicted that "machine won the game for G1–G7," and "the winning
rate difference between machine and human is the smallest when G6 was already
played 100 moves" [3].

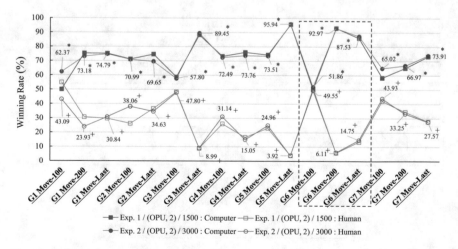

Fig. 15 Winning rate curves for G1 to G7 [3]

- **PFML-Based Prediction Agent**

The PFML-based prediction agent is to show the after-learning performance based on the PSO and FML. We used Master's 60 games in Jan. 2017 [23] as the training data of PFML learning mechanism and then chose three games from "The Future of Go Summit" [23], that is, Games 8–10 in Table 4, to test the learned performance. The desired outputs of the training data were given by extracting the domain experts' knowledge from the book "Move of Master" published in Taiwan [25]. We gave the desired output, including B^{++}, B^+, U, W^+, and W^{++} to a specific move and moves near its neighborhood by following Table 1. There are two machines installed DyNaDF Platform with FAIR Open Source DarkForest: including (1) NCHC (Tainan, Taiwan) with GeForce GTX 1080 × 4 and (2) NUTN (Tainan, Taiwan) with Quadro K2200 × 1 and Quadro M2000 × 1. We used these two machines to do the experiments for setting Number of simulations to 20,000 and 3000, respectively.

Figure 16a shows before-learning and after-learning accuracy when we sampled 3 or 4 sub-games from the game to infer its final game result [3], where BL and AL denote before learning and after learning, respectively; (2) MG1-5, MG1-10, MG1-40, and MG1-60 denote that PFML-based prediction agent learned the domain knowledge from Games 1-5, Games 1-10, Games 1-40, and Games 1-60 of Master

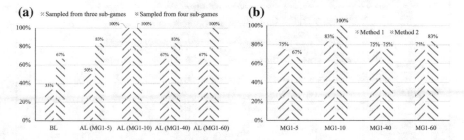

Fig. 16 a Before-learning and after-learning accuracy for different sub-game sample and **b** after-learning accuracy for different methods that give desired outputs

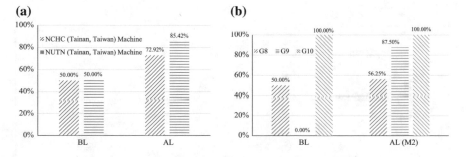

Fig. 17 Accuracy when compared with **a** different machines and **b** G8-G10

Table 5 Partial after-learning FML

```xml
<?xml version="1.0" encoding="UTF-8"?>
<fuzzySystem xmlns="http://www.ieee1855.org" name="GameSystem" networkAddress="127.0.0.1">
  <knowledgeBase networkAddress="127.0.0.1">
    <fuzzyVariable name="BSN" scale="" domainleft="0" domainright="20000" type="Input" accumulation="MAX" defuzzifier="COG" defaultValue="0.0" networkAddress="127.0.0.1">
      <fuzzyTerm name="Small" complement="false">
        <trapezoidShape param1="0" param2="0" param3="9466.12396529594" param4="14494.38778772"/>
      </fuzzyTerm>
      <fuzzyTerm name="Average" complement="false">
        <trapezoidShape param1="12789.4517683906" param2="15551.1595223206" param3="16910.6971562919" param4="19668.5142201235"/>
      </fuzzyTerm>
      <fuzzyTerm name="Big" complement="false">
        <trapezoidShape param1="18334.5948582598" param2="19814.2473572435" param3="20000" param4="20000"/>
      </fuzzyTerm>
    </fuzzyVariable>
    <fuzzyVariable name="WSN" scale="" domainleft="0" domainright="20000" type="Input" accumulation="MAX" defuzzifier="COG" defaultValue="0.0" networkAddress="127.0.0.1">
      <fuzzyTerm name="Small" complement="false">
        <trapezoidShape param1="0" param2="0" param3="443.243081216463" param4="11977.9189247817"/>
      </fuzzyTerm>
      <fuzzyTerm name="Average" complement="false">
        <trapezoidShape param1="3951.64636710093" param2="12067.8227114097" param3="12364.6131719719" param4="16810.3296457942"/>
      </fuzzyTerm>
      <fuzzyTerm name="Big" complement="false">
        <trapezoidShape param1="14118.9542353869" param2="19817.3772536807" param3="20000" param4="20000"/>
      </fuzzyTerm>
    </fuzzyVariable>
    <fuzzyVariable name="BWR" scale="" domainleft="0" domainright="1" type="Input" accumulation="MAX" defuzzifier="COG" defaultValue="0.0" networkAddress="127.0.0.1">
      <fuzzyTerm name="Small" complement="false">
        <trapezoidShape param1="0" param2="0" param3="0.0890039183288133" param4="0.431612383480871"/>
      </fuzzyTerm>
      <fuzzyTerm name="Average" complement="false">
        <trapezoidShape param1="0.363104316715857" param2="0.439693657054856" param3="0.445473452407203" param4="0.479785414053812"/>
      </fuzzyTerm>
      <fuzzyTerm name="Big" complement="false">
        <trapezoidShape param1="0.479322344954228" param2="0.505831223250701" param3="1" param4="1"/>
      </fuzzyTerm>
    </fuzzyVariable>
```
⋮

```xml
    <fuzzyVariable name="CGS" scale="" domainleft="0" domainright="10" type="Output" accumulation="MAX" defuzzifier="COG" defaultValue="0.0" networkAddress="127.0.0.1">
      <fuzzyTerm name="BlackObviousAdvantage" complement="false">
        <trapezoidShape param1="0" param2="0" param3="1.05950832016088" param4="2.20900177892658"/>
      </fuzzyTerm>
      <fuzzyTerm name="BlackPossibleAdvantage" complement="false">
        <trapezoidShape param1="1.06601939947709" param2="2.21613320978007" param3="2.24358282880617" param4="2.28690876817452"/>
      </fuzzyTerm>
      <fuzzyTerm name="UncertainSituation" complement="false">
        <trapezoidShape param1="2.24943127456175" param2="4.07676374020868" param3="5.10137682147647" param4="8.25169355616227"/>
      </fuzzyTerm>
      <fuzzyTerm name="WhitePossibleAdvantage" complement="false">
        <trapezoidShape param1="7.98279232644494" param2="8.35315682782079" param3="8.39292046435599" param4="9.10699596350135"/>
      </fuzzyTerm>
      <fuzzyTerm name="WhiteObviousAdvantage" complement="false">
        <trapezoidShape param1="8.51843457812179" param2="9.94722153726252" param3="10" param4="10"/>
      </fuzzyTerm>
    </fuzzyVariable>
  </knowledgeBase>
  <mamdaniRuleBase name="ruleBase1" activationMethod="MIN" andMethod="MIN" orMethod="MAX" networkAddress="127.0.0.1">
    <rule name="rule-1" andMethod="MIN" orMethod="MAX" connector="AND" weight="1.0" networkAddress="127.0.0.1">
      <antecedent>
        <clause>
          <variable>BSN</variable>
          <term>Small</term>
        </clause>
        <clause>
          <variable>WSN</variable>
          <term>Small</term>
        </clause>
        <clause>
          <variable>BWR</variable>
          <term>Small</term>
        </clause>
        <clause>
          <variable>WWR</variable>
          <term>Small</term>
        </clause>
        <clause>
          <variable>BSA</variable>
          <term>Average</term>
        </clause>
        <clause>
          <variable>WSA</variable>
          <term>Average</term>
        </clause>
      </antecedent>
      <consequent>
        <then>
          <clause>
            <variable>CGS</variable>
            <term>UncertainSituation</term>
          </clause>
        </then>
      </consequent>
    </rule>
```
⋮
```xml
  </mamdaniRuleBase>
</fuzzySystem>
```

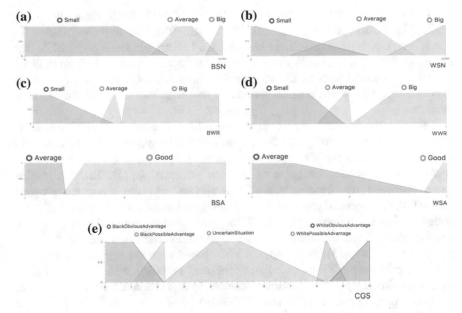

Fig. 18 After-learning fuzzy sets for fuzzy variables **a** *BSN*, **b** *WSN*, **c** *BWR*, **d** *WWR*, **e** *BSA* and **f** *WSA*

60 games in Jan. 2017, respectively. Observe Fig. 16a, it indicates that (1) after learning performs better than before learning and (2) sampling game's information from four sub-games to infer the game result performs better than sampling from three sub-games. Figure 16b shows that (1) learning the domain knowledge from MG1-10 and MG1-60 gets the first and the second best performances, respectively, and (2) on average, adopting Method 2 to giving desired output gets the better performance than Method 1. Figure 17a shows that after learning, NUTN (Tainan, Taiwan) machine gets better performance between NCHC (Tainan, Taiwan) one. Figure 17b shows after learning, G9 has an impressive performance compared to the other two games. Table 5 shows partial after-learning FML when we learned the domain knowledge from MG1-10 and gave the desired outputs based on Method 2 listed in Table 1. Figures 18a–f show the after-learning fuzzy sets for fuzzy variables *BSN*, *WSN*, *BWR*, *WWR*, *BSA*, and *WSA*, respectively, when we adopted the proposed PFML learning mechanism to learn 3000 generations with 20 particles.

5 Conclusions

Since 2009, IEEE Computational Intelligence Society (CIS) has helped to fund human vs. computer Go competitions in the past IEEE CIS-flag conferences, including the FUZZ-IEEE 2009, IEEE WCCI 2010, IEEE SSCI 2011, FUZZ-IEEE

2011, IEEE WCCI 2012, FUZZ-IEEE 2013, FUZZ-IEEE 2015, IEEE CIG 2015, IEEE WCCI 2016, and FUZZ-IEEE 2017. IEEE SMC Society also funded the special event of Human and Smart Machine Co-Learning in IEEE SMC 2017. The handicaps for 19×19 board game have been decreased from 29 to 0 up to 2016. In 2017, powerful computer Go programs such as AlphaGo and DeepZenGo are expected to give top professional humans a few handicap stones to make for an even match.

This chapter describes the research from T2 FS-based MoGoTW system to Dynamic DarkForest (DyNaDF) for human and machine co-learning on Go. A T2 FS-based adaptive linguistic assessment system for semantic analysis and human performance evaluation on game of Go was presented in [1]. Through playing games between the invited human Go players and the computer program MoGoTW, the proposed approach in [1] inferred the human's rank according to the collected simulation number of MCTS, the game's komi setting, the winning rate, number of consecutive winning games, and so on. In addition, a robotic prediction agent was proposed [3]. The proposed FML-based decision support engine computed the winning possibility based on DarkForest's prediction and the partial game situation inferred by FML assessment engine. In this chapter, we further combine FML and particle swarm optimization (PSO) to learn the domain knowledge of Master's 60 games [23]. After learning, we use the proposed approaches in [3] to infer the game results of the "Future of Go Summit in Wuzhen" in May 2017.

The experimental results show some conclusions as follows: (1) Before learning, the proposed T2 FS-based adaptive linguistic assessment system performs better than T1 FS-based one. (2) For the proposed T2 FS-based adaptive linguistic assessment system, its after-learning results are better than its own before-learning ones. (3) The type-2 system has a better ability than the type-1 system to handle the real-world applications with a high-level uncertainty [5] even though T2 FS is not always suitable for all of the situations. (4) Sampling game's information from four sub-games to infer the game result performs better than sampling from three sub-games. (5) On average, adopting Method 2 listed in Table 1 to giving desired output gets the better performance than Method 1. In the future, we will combine the robot to learn together with Go players, include more data to validate the learning performance, and generate play-comments on a game to highlight the positional strategic plan followed by a player during a sequence of moves.

Acknowledgements The authors would like to thank the Ministry of Science Technology (MOST) of Taiwan for partially supporting this international cooperation research project under the grant MOST 105-2221-E-024-017, MOST 105-2622-E-024-003-CC2, and MOST 106-3114-E-024-001. Additionally, the authors would like to thank all invited Go players from Haifong Weiqi Academy, Taiwan for their kind help and also thank Dr. Yuandong Tian and Yan Zhu from Facebook AI Research (FAIR). Finally, the authors would like to thank the National Center of High Performance Computing (NCHC) and Kaohsiung City Government, Taiwan for their computing resource support.

References

1. C.S. Lee, M.H. Wang, M.J. Wu, O. Teytaud, S.J. Yen, T2FS-based adaptive linguistic assessment system for semantic analysis and human performance evaluation on game of Go. IEEE Trans. Fuzzy Syst. **23**(2), 400–420 (2015)
2. C.S. Lee, M.H. Wang, S.J. Yen, T.H. Wei, I.C. Wu, P.C. Chou, C.H. Chou, M.W. Wang, T. H. Yang, Human vs. computer Go: Review and prospect. IEEE Comput. Intell. Mag. **11**(3), 67–72 (2016)
3. C.S. Lee, M.H. Wang, C.H. Kao, S.C. Yang, Y. Nojima, R. Saga, N. Shuo, N. Kubota, FML-based prediction agent and its application to game of Go. Joint 17th World congress of international fuzzy systems association and 9th international conference on soft computing and intelligent systems (IFSA-SCIS 2017), Otsu, Japan, 27–30 June 2017
4. J.M. Mendel, *Uncertain Rule-Based Fuzzy Logic Systems: Introduction and New directions* (Prentice-Hall, Upper Saddle River, NJ, 2001)
5. A. Sadeghian, J.M. Mendel, H. Tahayori, *Advances in Type-2 Fuzzy Sets: Theory and Applications*. (Springer, New York, 2013)
6. J.M. Mendel, On KM algorithms for solving type-2 fuzzy set problems. IEEE Trans. Fuzzy Syst. **21**(3), 426–446 (2013)
7. H. Hagras, C. Wagner, Towards the wide speard use of type-2 fuzzy logic systems in real world applications. Comput. Intell. Mag. **7**(3), 14–24 (2012)
8. G. Acampora, C.S. Lee, M.H. Wang, V. Loia, *On the Power of Fuzzy Markup Language* (Springer-Verlag, Germany, 2013)
9. C.S. Lee, M.H. Wang, G. Acampora, C.Y. Hsu, H. Hagras, Diet assessment based on type-2 fuzzy ontology and fuzzy markup language. Int. J. Intell. Syst. **25**(12), 1187–1216 (2010)
10. IEEE Computational Intelligence Society, 1855–2016-IEEE Standard for Fuzzy Markup Language (2016). [Online] Available: http://ieeexplore.ieee.org/servlet/opac?punumber= 7479439
11. E. van der Werf, *AI techniques for the game of Go. Datawyse b.v* (Maastricht, The Netherlands, 2004)
12. M. Mueller, Computer Go. Artif. Intell. **134**(1–2), 145–179 (2002)
13. Google Deep Mind, AlphaGo at The Future of Go Summit (2017), 23–27 May 2017. [Online] Available: https://deepmind.com/research/alphago/alphago-china/
14. D. Silver, A. Huang, C.J. Maddison, A. Guez, L. Sifre, G. van den Driessche, J. Schrittwieser, I. Antonoglou, V. Panneershelvam, M. Lantot, S. Dieleman, D. Grewe, J. Nham, N. Kalchbrenner, I. Sutskever, T. Lillicrap, M. Leach, K. Kavukcuoglu, T. Graepel, D. Hassabis, Mastering the game of Go with deep neural networks and tree search. Nature **529**, 484–489 (2016)
15. C.B. Browne, E. Powley, D. Whitehouse, S.M. Lucas, P.I. Cowling, P. Rohlfshagen, S. Tavener, D. Perez, S. Samothrakis, S. Colton, A survey of Monte Carlo tree search methods. IEEE Trans. Comput. Intell. AI Games **4**(1), 1–43 (2012)
16. X. Cai, G.K. Venayagamoorthy, D.C. Wunsch II, Evolutionary swarm neural network game engine for Capture Go. Neural Networks **23**(2), 295–305 (2010)
17. S. Gelly, D. Silver, Monte-Carlo tree search and rapid action value estimation in computer Go. Artif. Intell. **175**(11), 1856–1875 (2011)
18. Y.D. Tian, Y. Zhu, Better computer Go player with neural network and long-term prediction (2016). [Online] Available: https://arxiv.org/abs/1511.06410
19. Y.D. Tian, Facebook research/darkforest Go. (2016). [Online] Available: https://github.com/ facebookresearch/darkforestGo
20. C. Clark, A. Storkey, Teaching deep convolutional neural networks to play Go (2014). [Online] Available: http://arxiv.org/abs/1412.3409
21. C.J. Maddison, A. Huang, I. Sutskever, D. Silver, Move evaluation in Go using deep convolutional neural networks (2014). [Online] Available: http://arxiv.org/abs/1412.6564

22. C.S. Lee, M.H. Wang, S.C. Yang, P.H. Hung, S.W. Lin, N. Shuo, N. Kubota, C.H. Chou, P. C. Chou, C.H. Kao, FML-based dynamic assessment agent for human-machine cooperative system on game of Go. Int. J. Uncertainty Fuzziness Knowl. Based Syst. (IJUFKS). **25**(5), 677–705 (2017)
23. AlphaGo (2017). [Online] Available: http://homepages.cwi.nl/~aeb/go/games/games/AlphaGo/
24. C.S. Lee, M.H. Wang, C.S. Wang, O. Teytaud, J. Liu, S.W. Lin, P.H. Hung, PSO-based fuzzy markup language for student learning performance evaluation and educational application. IEEE Trans. Fuzzy Syst. (2017) (Revised)
25. D.M. Hsia, C.H. Hsiao, S.Y. Lin, *Move of Master* (Ming Chu Press, Taiwan, 2017). (in Chinese)
26. C.S. Lee, M.H. Wang, O. Teytaud, Fuzzy ontologies for the game of Go, in *On Fuzziness. A Homage to Lotfi A. Zadeh*, ed. by R. Seising, E. Trillas, C. Moraga, S. Termini (Springer, Berlin, New York, 2013), pp. 367–372
27. C.S. Lee, M.H. Wang, M.J. Wu, Y. Nakagawa, H. Tsuji, Y. Yamazaki, K. Hirota, Soft-computing-based emotional expression mechanism for game of Computer Go. Soft. Comput. **17**(7), 1263–1282 (2013)
28. C.S. Lee, M.H. Wang, Y.J. Chen, H. Hagras, M.J. Wu, O. Teytaud, Genetic fuzzy markup language for game of NoGo. Knowl. Based Syst. **34**, 64–80 (2012)
29. C.S. Lee, M.H. Wang, C. Chaslot, J.B. Hoock, A. Rimmel, O. Teytaud, S.R. Tsai, S.C. Hsu, T.P. Hong, The computational intelligence of MoGo revealed in Taiwan's computer Go tournaments. IEEE Tran. Comput. Intell. AI Games **1**(1), 73–89 (2009)
30. J.B. Hoock, C.S. Lee, A. Rimmel, F. Teytaud, M.H. Wang, O. Teytaud, Intelligent agents for the game of Go. IEEE Comput. Intell. Mag. **5**(4), 28–42 (2010)
31. S.E. Embretson, S.P. Reise, *Item Response Theory* (Taylor & Francis, 2000)

Ordered Novel Weighted Averages

Dongrui Wu and Jian Huang

Abstract The novel weighted averages (NWAs) are extensions of the linear arithmetic weighted average and are powerful tools in aggregating diverse inputs including numbers, intervals, type-1 fuzzy sets (T1 FSs), words modeled by interval type-2 fuzzy sets, or a mixture of them. In contrast to the linear arithmetic weighted average, the ordered weighed average (OWA) is a nonlinear operator that can implement more flexible mappings, and hence it has been widely used in decision-making. In many situations, however, providing crisp numbers for either the sub-criteria or the weights is problematic (there could be uncertainties about them), and it is more meaningful to provide intervals, T1 FSs, words, or a mixture of all of these, for the sub-criteria and weights. Ordered NWAs are introduced in this chapter. They are also compared with NWAs and Zhou et al's fuzzy extensions of the OWA. Examples show that generally the three aggregation operators give different results.

1 Introduction

The *weighted average* (WA) is arguably the earliest and still most widely used form of aggregation or fusion. We remind the reader of the well-known formula for the WA, i.e.,

$$y = \frac{\sum_{i=1}^{n} x_i w_i}{\sum_{i=1}^{n} w_i}, \tag{1}$$

D. Wu (✉) · J. Huang
School of Automation, Huazhong University of Science and Technology,
Wuhan, Hubei, China
e-mail: drwu@hust.edu.cn

J. Huang
e-mail: huang_jan@mail.hust.edu.cn

© Springer International Publishing AG 2018
R. John et al. (eds.), *Type-2 Fuzzy Logic and Systems*,
Studies in Fuzziness and Soft Computing 362,
https://doi.org/10.1007/978-3-319-72892-6_2

25

in which w_i are the weights (real numbers) that act upon the sub-criteria x_i (real numbers). In this chapter, the term *sub-criteria* can mean data, features, decisions, recommendations, judgments, scores, etc. In (1), normalization is achieved by dividing the weighted numerator sum by the sum of all of the weights.

The arithmetic WA (AWA) is the one we are all familiar with and is the one in which all sub-criteria and weights in (1) are real numbers. In many situations [1–7], however, providing crisp numbers for either the sub-criteria or the weights is problematic (there could be uncertainties about them), and it is more meaningful to provide intervals, type-1 fuzzy sets (T1 FSs), words modeled by interval type-2 fuzzy sets (IT2 FSs), or a mixture of all of these, for the sub-criteria and weights. The resulting WAs are called novel weighted averages (NWAs), which have been introduced in [2, 3, 6].

The *ordered weighted average* (OWA) operator [8–16], a generalization of the linear WA operator, was proposed by Yager to aggregate experts' opinions in decision making:

Definition 1 An OWA operator of dimension n is a mapping $y_{OWA} : R^n \to R$, which has an associated set of weights $\mathbf{w} = \{w_1, \dots, w_n\}$ for which $w_i \in [0, 1]$, i.e.,

$$y_{OWA} = \frac{\sum_{i=1}^{n} w_i x_{\sigma(i)}}{\sum_{i=1}^{n} w_i} \tag{2}$$

where $\sigma : \{1, \dots, n\} \to \{1, \dots, n\}$ is a permutation function such that $\{x_{\sigma(1)}, x_{\sigma(2)}, \dots, x_{\sigma(n)}\}$ are in descending order. ∎

The key feature of the OWA operator is the ordering of the sub-criteria by value, a process that introduces a nonlinearity into the operation. It can be shown that the OWA operator is in the class of mean operators [17] as it is commutative, monotonic, and idempotent. It is also easy to see that for any $\mathbf{w}$, $\min_i x_i \leq y_{OWA} \leq \max_i x_i$.

The most attractive feature of the OWA operator is that it can implement different aggregation operators by setting the weights differently [8], e.g., by setting $w_i = 1/n$ it implements the mean operator, by setting $w_1 = 1$ and $w_i = 0$ $(i = 2, \dots, n)$ it implements the maximum operator, by setting $w_i = 0$ $(i = 1, \dots, n - 1)$ and $w_n = 1$ it implements the minimum operator, and by setting $w_1 = w_n = 0$ and $w_i = 1/(n - 1)$ it implements the so-called olympic aggregator, which is often used in obtaining aggregated scores from judge in olympic events such as gymnastics and diving.

Yager's original OWA operator [12] considers only crisp numbers. Again, in many situations, it is more meaningful to provide intervals, T1 FSs, words modeled by IT2 FSs, or a mixture of all of these, for the sub-criteria and weights. Ordered NWAs (ONWAs) are the focus of this chapter.

The rest of this chapter is organized as follows: Sect. 2 introduces the NWAs. Section 3 proposes ONWAs. Section 4 compares ONWAs with NWAs and Zhou et al's fuzzy extensions of the OWA. Finally, conclusions are drawn in Sect. 5.

2 Novel Weighted Averages (NWAs)

Definition 2 A NWA is a WA in which at least one sub-criterion or weight is not a single real number, but instead is an interval, T1 FS or an IT2 FS, in which case such sub-criteria or weights are called *novel models*. ∎

How to compute (1) for these novel models is described in this section. Because there can be four possible models for sub-criteria or weights, there can be 16 different WAs, as summarized in Fig. 1.

Definition 3 When at least one sub-criterion or weight is modeled as an interval, and all other sub-criteria or weights are modeled by no more than such a model, the resulting WA is called an *Interval WA* (IWA). ∎

Definition 4 When at least one sub-criterion or weight is modeled as a T1 FS, and all other sub-criteria or weights are modeled by no more than such a model, the resulting WA is called a *Fuzzy WA* (FWA). ∎

Definition 5 When at least one sub-criterion or weight is modeled as an IT2 FS, the resulting WA is called a *Linguistic WA* (LWA). ∎

Definition 2 (Continued) By a NWA is meant an IWA, FWA or LWA. ∎

In order to reduce the number of possible derivations from 15 (the AWA is excluded) to three, it is assumed that: for the IWA *all* sub-criteria and weights are modeled as intervals, for the FWA *all* sub-criteria and weights are modeled as T1 FSs, and for the LWA *all* sub-criteria and weights are modeled as IT2 FSs.

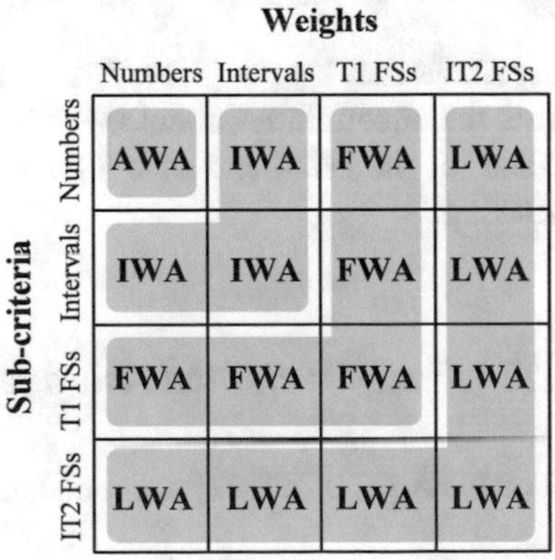

Fig. 1 Matrix of possibilities for a WA

		Weights		
		Numbers	Intervals	T1 FSs	IT2 FSs
Sub-criteria	Numbers	AWA	IWA	FWA	LWA
	Intervals	IWA	IWA	FWA	LWA
	T1 FSs	FWA	FWA	FWA	LWA
	IT2 FSs	LWA	LWA	LWA	LWA

2.1 Interval Weighted Average (IWA)

The IWA is defined as:

$$Y_{IWA} \equiv \frac{\sum_{i=1}^{n} X_i W_i}{\sum_{i=1}^{n} W_i} = [l, r] \tag{3}$$

where

$$X_i = [a_i, \; b_i] \quad i = 1, ..., n \tag{4}$$

$$W_i = [c_i, \; d_i] \quad i = 1, ..., n \tag{5}$$

and Y_{IWA} is also an interval completely determined by its two end-points l and r,

$$l = \min_{\substack{x_i \in X_i \\ w_i \in W_i}} \frac{\sum_{i=1}^{n} x_i w_i}{\sum_{i=1}^{n} w_i} = \min_{w_i \in W_i} \frac{\sum_{i=1}^{n} a_i w_i}{\sum_{i=1}^{n} w_i} \tag{6}$$

$$r = \max_{\substack{x_i \in X_i \\ w_i \in W_i}} \frac{\sum_{i=1}^{n} x_i w_i}{\sum_{i=1}^{n} w_i} = \max_{w_i \in W_i} \frac{\sum_{i=1}^{n} b_i w_i}{\sum_{i=1}^{n} w_i} \tag{7}$$

and they can easily be computed by the KM or EKM Algorithms [18–21].

Example 1 Suppose for $n = 5$, $\{x_i\}|_{i=1,...,5} = \{5, 7.5, 7, 6.5, 2\}$ and $\{w_i\}|_{i=1,...,5} = \{4, 2.5, 8, 1.8, 6\}$, so that the arithmetic WA $y_{AWA} = 5.31$. Let λ denote any of these crisp numbers. In this example, for the IWA, $\lambda \to [\lambda - \delta, \lambda + \delta]$, where δ may be different for different λ, i.e.,

$$\{x_i\}|_{i=1,...,5} \to \{[4.5, 5.5], [7.0, 8.0], [4.2, 9.8], [6.0, 7.0], [1.0, 3.0]\}$$

$$\{w_i\}|_{i=1,...,5} \to \{[2.8, 5.2], [2.0, 3.0], [7.6, 8.4], [0.9, 2.7], [5.0, 7.0]\}.$$

It follows that $Y_{IWA} = [3.49, 7.12]$. The important difference between y_{AWA} and Y_{IWA} is that the uncertainties about the sub-criteria and weights have led to an uncertainty band for the IWA, and such a band may play a useful role in subsequent decision making. ∎

2.2 Fuzzy Weighted Average (FWA)

The FWA [2, 3, 22–27] is defined as:

$$Y_{FWA} \equiv \frac{\sum_{i=1}^{n} X_i W_i}{\sum_{i=1}^{n} W_i} \tag{8}$$

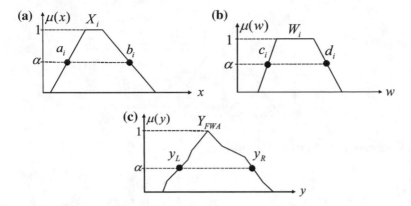

Fig. 2 α-cuts on **a** X_i, **b** W_i, and, **c** Y_{FWA}

where X_i and W_i are T1 FSs, and Y_{FWA} is also a T1 FS. Note that (8) is an *expressive* way to represent the FWA because it is not computed using multiplications, additions and divisions, as expressed by it. Instead, it has been shown [2–4, 27] that the FWA can be computed by using the α-cut Decomposition Theorem [28], where each α-cut on Y_{FWA} is an IWA of the corresponding α-cuts on X_i and W_i.

Denote the α-cut on Y_{FWA} as $[y_L(\alpha), y_R(\alpha)]$, and the α-cut on X_i and W_i as $[a_i(\alpha), b_i(\alpha)]$ and $[c_i(\alpha), d_i(\alpha)]$, respectively, as shown in Fig. 2. Then [2–4, 27],

$$y_L(\alpha) = \min_{\forall w_i(\alpha) \in [c_i(\alpha), d_i(\alpha)]} \frac{\sum_{i=1}^{n} a_i(\alpha) w_i(\alpha)}{\sum_{i=1}^{n} w_i(\alpha)} \tag{9}$$

$$y_R(\alpha) = \max_{\forall w_i(\alpha) \in [c_i(\alpha), d_i(\alpha)]} \frac{\sum_{i=1}^{n} b_i(\alpha) w_i(\alpha)}{\sum_{i=1}^{n} w_i(\alpha)} \tag{10}$$

$y_L(\alpha)$ and $y_R(\alpha)$ can be computed by the KM or EKM algorithms.

The following algorithm is used to compute Y_{FWA}:

1. For each $\alpha \in [0, 1]$, the corresponding α-cuts of the T1 FSs X_i and W_i are first computed, i.e., compute

$$X_i(\alpha) = [a_i(\alpha), b_i(\alpha)] \quad i = 1, ..., n \tag{11}$$
$$W_i(\alpha) = [c_i(\alpha), d_i(\alpha)] \quad i = 1, ..., n \tag{12}$$

2. For each $\alpha \in [0, 1]$, compute $y_L(\alpha)$ in (9) and $y_R(\alpha)$ in (10) using the KM or EKM Algorithms.
3. Connect all left-coordinates $(y_L(\alpha), \alpha)$ and all right-coordinates $(y_R(\alpha), \alpha)$ to form the T1 FS Y_{FWA}.

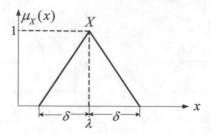

Fig. 3 Illustration of a T1 FS used in Example 2

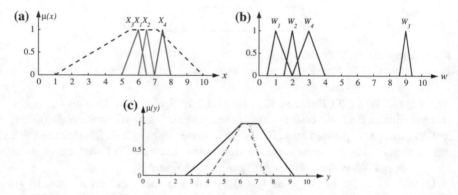

Fig. 4 Example 2: **a** X_i, **b** W_i, and, **c** Y_{FWA} (solid curve), Y_{OFWA} (dashed curve) and Y_{T1FOWA} (dotted curve)

Example 2 This is a continuation of Example 1 in which each interval is assigned a symmetric triangular T1 FS that is centered at the mid-point (λ) of the interval, has membership grade equal to one at that point, and is zero at the interval end-points ($\lambda - \delta$ and $\lambda + \delta$) (see the triangle in Fig. 3). The resulting X_i and W_i are plotted in Fig. 4a and Fig. 4b, respectively. The FWA is depicted in Fig. 4c as the solid curve. Although Y_{FWA} appears to be triangular, its sides are actually slightly curved.

The support of Y_{FWA} is [3.49, 7.12], which is the same as Y_{IWA} (see Example 1). This will always occur because the support of Y_{FWA} is the $\alpha = 0$ α-cut, and this is Y_{IWA}. The T1 FS Y_{FWA} indicates that more emphasis should be given to values of variable y that are closer to its apex, whereas the interval Y_{IWA} indicates that equal emphasis should be given to all values of variable y in its interval. The former reflects the propagation of the non-uniform uncertainties through the FWA, and can be used in future decisions.   ■

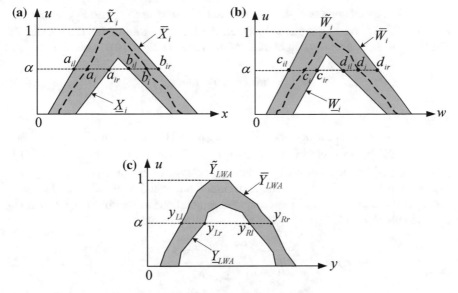

Fig. 5 **a** $\tilde{X}_i$ and an α-cut, **b** $\tilde{W}_i$ and an α-cut, and, **c** $\tilde{Y}_{LWA}$ and an α-cut

2.3 Linguistic Weighted Average (LWA)

The LWA is defined as:

$$\tilde{Y}_{LWA} \equiv \frac{\sum_{i=1}^{n} \tilde{X}_i \tilde{W}_i}{\sum_{i=1}^{n} \tilde{W}_i} \tag{13}$$

where $\tilde{X}_i$ and $\tilde{W}_i$ are IT2 FSs, and $\tilde{Y}_{FWA}$ is also an IT2 FS. Again, (13) is an *expressive* way to describe the LWA. To compute $\tilde{Y}_{LWA}$, one only needs to compute its LMF $\underline{Y}_{LWA}$ and UMF $\bar{Y}_{LWA}$.

Let W_i be an embedded T1 FS [19] of $\tilde{W}_i$, as shown in Fig. 5b. Because in (13) $\tilde{X}_i$ only appears in the numerator of $\tilde{Y}_{LWA}$, it follows that

$$\underline{Y}_{LWA} = \min_{\forall W_i \in [\underline{W}_i, \bar{W}_i]} \frac{\sum_{i=1}^{n} \underline{X}_i W_i}{\sum_{i=1}^{n} W_i} \tag{14}$$

$$\bar{Y}_{LWA} = \max_{\forall W_i \in [\underline{W}_i, \bar{W}_i]} \frac{\sum_{i=1}^{n} \bar{X}_i W_i}{\sum_{i=1}^{n} W_i} \tag{15}$$

The α-cut based approach [4, 5] is also used to compute $\underline{Y}_{LWA}$ and $\bar{Y}_{LWA}$. First, the heights of $\underline{Y}_{LWA}$ and $\bar{Y}_{LWA}$ need to be determined. Because all UMFs are normal T1 FSs, $h_{\bar{Y}_{LWA}} = 1$. Denote the height of $\underline{X}_i$ as $h_{\underline{X}_i}$ and the height of $\underline{W}_i$ as $h_{\underline{W}_i}$. Let

$$h_{\min} = \min\{\min_{\forall i} h_{\underline{X}_i}, \min_{\forall i} h_{\underline{W}_i}\} \tag{16}$$

Then [5], $h_{\underline{Y}_{LWA}} = h_{\min}$.

Let $[a_{il}(\alpha), b_{ir}(\alpha)]$ be an α-cut on $\bar{X}_i$, $[a_{ir}(\alpha), b_{il}(\alpha)]$ be an α-cut on $\underline{X}_i$ [see Fig. 5a], $[c_{il}(\alpha), d_{ir}(\alpha)]$ be an α-cut on $\bar{W}_i$, $[c_{ir}(\alpha), d_{il}(\alpha)]$ be an α-cut on $\underline{W}_i$ [see Fig. 5b], $[y_{Ll}(\alpha), y_{Rr}(\alpha)]$ be an α-cut on $\bar{Y}_{LWA}$, and $[y_{Lr}(\alpha), y_{Rl}(\alpha)]$ be an α-cut on $\underline{Y}_{LWA}$ [see Fig. 5c], where the subscripts l and L mean *left* and r and R mean *right*. The endpoints of the α-cuts on $\tilde{Y}_{LWA}$ are computed as solutions to the following four optimization problems [4, 5]:

$$y_{Ll}(\alpha) = \min_{\forall w_i \in [c_{il}(\alpha), d_{ir}(\alpha)]} \frac{\sum_{i=1}^n a_{il}(\alpha) w_i}{\sum_{i=1}^n w_i}, \quad \alpha \in [0, 1] \tag{17}$$

$$y_{Rr}(\alpha) = \max_{\forall w_i \in [c_{il}(\alpha), d_{ir}(\alpha)]} \frac{\sum_{i=1}^n b_{ir}(\alpha) w_i}{\sum_{i=1}^n w_i}, \quad \alpha \in [0, 1] \tag{18}$$

$$y_{Lr}(\alpha) = \min_{\forall w_i \in [c_{ir}(\alpha), d_{il}(\alpha)]} \frac{\sum_{i=1}^n a_{ir}(\alpha) w_i}{\sum_{i=1}^n w_i}, \quad \alpha \in [0, h_{\min}] \tag{19}$$

$$y_{Rl}(\alpha) = \max_{\forall w_i \in [c_{ir}(\alpha), d_{il}(\alpha)]} \frac{\sum_{i=1}^n b_{il}(\alpha) w_i}{\sum_{i=1}^n w_i}, \quad \alpha \in [0, h_{\min}] \tag{20}$$

(17)–(20) are again computed by the KM or EKM Algorithms.

Observe from (17), (18), and Figs. 5a and b that $y_{Ll}(\alpha)$ and $y_{Rr}(\alpha)$ only depend on the UMFs of $\tilde{X}_i$ and $\tilde{W}_i$, i.e., they are only computed from the corresponding α-cuts on the UMFs of $\tilde{X}_i$ and $\tilde{W}_i$; so,

$$\bar{Y}_{LWA} = \frac{\sum_{i=1}^n \bar{X}_i \bar{W}_i}{\sum_{i=1}^n \bar{W}_i}. \tag{21}$$

Because all $\bar{X}_i$ and $\bar{W}_i$ are normal T1 FSs, $\bar{Y}_{LWA}$ is also normal. The algorithm for computing $\bar{Y}_{LWA}$ is:

1. Select appropriate m α-cuts for $\overline{Y}_{LWA}$ (e.g., divide [0, 1] into $m - 1$ intervals and set $\alpha_j = (j - 1)/(m - 1), j = 1, 2, ..., m$).
2. For each α_j, find the corresponding α-cuts $[a_{il}(\alpha_j), b_{ir}(\alpha_j)]$ and $[c_{il}(\alpha_j), d_{ir}(\alpha_j)]$ on $\overline{X}_i$ and $\overline{W}_i$ ($i = 1, ..., n$). Use a KM or EKM algorithm to find $y_{Ll}(\alpha_j)$ in (17) and $y_{Rr}(\alpha_j)$ in (18).
3. Connect all left-coordinates $(y_{Ll}(\alpha_j), \alpha_j)$ and all right-coordinates $(y_{Rr}(\alpha_j), \alpha_j)$ to form the T1 FS $\overline{Y}_{LWA}$.

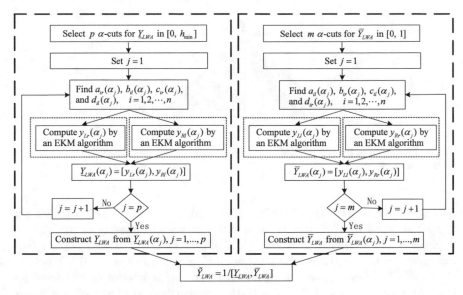

Fig. 6 A flowchart for computing the LWA [2, 3, 5]

Similarly, observe from (19), (20), and Fig. 5a and b that $y_{Lr}(\alpha)$ and $y_{Rl}(\alpha)$ only depend on the LMFs of $\tilde{X}_i$ and $\tilde{W}_i$; hence,

$$\underline{Y}_{LWA} = \frac{\sum_{i=1}^{n} \underline{X}_i \underline{W}_i}{\sum_{i=1}^{n} \underline{W}_i}. \tag{22}$$

Unlike $\bar{Y}_{LWA}$, which is a normal T1 FS, the height of $\underline{Y}_{LWA}$ is $h_{\min}$, the minimum height of all $\underline{X}_i$ and $\underline{W}_i$. The algorithm for computing $\underline{Y}_{LWA}$ is:

1. Determine $h_{\underline{X}_i}$ and $h_{\underline{W}_i}$, $i = 1, \dots, n$, and $h_{\min}$ in (16).
2. Select appropriate p α-cuts for $\underline{Y}_{LWA}$ (e.g., divide $[0, h_{\min}]$ into $p - 1$ intervals and set $\alpha_j = h_{\min}(j - 1)/(p - 1), j = 1, 2, \dots, p$).
3. For each α_j, find the corresponding α-cuts $[a_{ir}(\alpha_j), b_{il}(\alpha_j)]$ and $[c_{ir}(\alpha_j), d_{il}(\alpha_j)]$ on $\underline{X}_i$ and $\underline{W}_i$. Use a KM or EKM algorithm to find $y_{Lr}(\alpha_j)$ in (19) and $y_{Rl}(\alpha_j)$ in (20).
4. Connect all left-coordinates $(y_{Lr}(\alpha_j), \alpha_j)$ and all right-coordinates $(y_{Rl}(\alpha_j), \alpha_j)$ to form the T1 FS $\underline{Y}_{LWA}$.

In summary, computing $\tilde{Y}_{LWA}$ is equivalent to computing two FWAs, $\bar{Y}_{LWA}$ and $\underline{Y}_{LWA}$. A flowchart for computing $\underline{Y}_{LWA}$ and $\bar{Y}_{LWA}$ is given in Fig. 6. For triangular or trapezoidal IT2 FSs, it is possible to reduce the number of α-cuts for both $\underline{Y}_{LWA}$ and $\bar{Y}_{LWA}$ by choosing them only at *turning points*, i.e., points on the LMFs and UMFs of X_i and W_i ($i = 1, 2, \dots, n$) at which the slope of these functions changes.

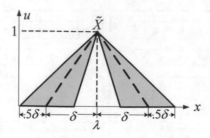

Fig. 7 Illustration of the IT2 FS used in Example 3. The dashed lines are the corresponding T1 FS used in Example 2

Example 3 This is a continuation of Example 2 where each sub-criterion and weight is now assigned an FOU that is for a 50% symmetrical blurring of the T1 MF depicted in Fig. 3 (see Fig. 7). The left half of each FOU has support on the x (w)-axis given by the interval of real numbers $[(\lambda - \delta) - 0.5\delta, (\lambda - \delta) + 0.5\delta]$ and the right-half FOU has support on the x-axis given by the interval of real numbers $[(\lambda + \delta) - 0.5\delta, (\lambda + \delta) + 0.5\delta]$. The UMF is a triangle defined by the three points $(\lambda - \delta - 0.5\delta, 0), (\lambda, 1), (\lambda + \delta + 0.5\delta, 0)$, and the LMF is a triangle defined by the three points $(\lambda - \delta + 0.5\delta, 0), (\lambda, 1), (\lambda + \delta - 0.5\delta, 0)$. The resulting sub-criterion and weight FOUs are depicted in Figs. 8a and b, respectively, and $\tilde{Y}_{LWA}$ is depicted in Fig. 8c as the solid curve. Although $\tilde{Y}_{LWA}$ appears to be symmetrical, it is not.

Comparing Figs. 8c and 4c, observe that $\tilde{Y}_{LWA}$ is spread out over a larger range of values than is Y_{FWA}, reflecting the additional uncertainties in the LWA due to the blurring of sub-criteria and weights. This information can be used in future decisions.

Another way to interpret $\tilde{Y}_{LWA}$ is to associate values of y that have the largest vertical intervals (i.e., primary memberships) with values of greatest uncertainty; hence, there is no uncertainty at the three vertices of the UMF, and, e.g., for the right-half of $\tilde{Y}_{LWA}$ uncertainty increases from the apex of the UMF reaching its largest value at the right vertex of the LMF and then decreases to zero at the right vertex of the UMF. ∎

3 Ordered Novel Weighted Averages (ONWAs)

ONWAs, including ordered IWAs, ordered FWAs and ordered LWAs, are proposed in this section.

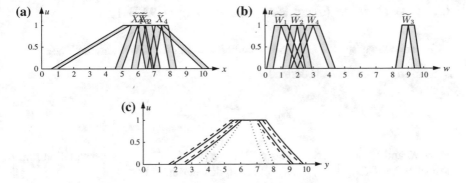

Fig. 8 Example 3: **a** $\tilde{X}_i$, **b** $\tilde{W}_i$, and, **c** $\tilde{Y}_{LWA}$ (solid curve), $\tilde{Y}_{OLWA}$ (dashed curve) and $\tilde{Y}_{IT2FOWA}$ (dotted curve)

3.1 The Ordered Interval Weighted Average (OIWA)

As its name suggests, the OIWA is a combination of the OWA and the IWA.

Definition 6 An OIWA is defined as

$$Y_{OIWA} = \frac{\sum_{i=1}^{n} W_i X_{\sigma(i)}}{\sum_{i=1}^{n} W_{\sigma(i)}} \tag{23}$$

where X_I and W_I are intervals defined in (4) and (5), respectively, and $\sigma : \{1, \ldots, n\} \rightarrow \{1, \ldots, n\}$ is a permutation function such that $\{X_{\sigma(1)}, X_{\sigma(2)}, \ldots, X_{\sigma(n)}\}$ are in descending order. ∎

Definition 7 A group of intervals $\{X_i\}_{i=1}^{n}$ are in descending order if $X_i \geq X_j$ for $\forall i < j$ by a ranking method. ∎

Any interval ranking method can be used to find σ. In this chapter, we first compute the center of each interval and then rank them to obtain the order of the corresponding intervals. This is a special case of Yager's first method [29] for ranking T1 FSs, where the T1 FSs degrade to intervals.

To compute Y_{OIWA}, we first sort X_i in descending order and call them by the same name, but now $X_1 \geq X_2 \geq \cdots \geq X_n$ (W_i are not changed during this step); then, the OIWA becomes an IWA.

Example 4 For the same crisp x_i and w_i used in Example 1, the OWA $y_{OWA} = 5.40$, which is different from $y_{AWA} = 5.31$. For the same interval X_i and W_i used in Example 1, the OIWA $Y_{OIWA} = [4.17, 6.66]$, which is different from $Y_{IWA} = [3.49, 7.12]$. ∎

3.2 The Ordered Fuzzy Weighted Average (OFWA)

As its name suggests, the OFWA is a combination of the OWA and the FWA.

Definition 8 An OFWA is defined as

$$Y_{OFWA} = \frac{\sum_{i=1}^{n} W_i X_{\sigma(i)}}{\sum_{i=1}^{n} W_{\sigma(i)}} \tag{24}$$

where X_i and W_i are T1 FSs, and $\sigma : \{1, \ldots, n\} \to \{1, \ldots, n\}$ is a permutation function such that $\{X_{\sigma(1)}, X_{\sigma(2)}, \ldots, X_{\sigma(n)}\}$ are in descending order. ∎

Definition 9 A group of T1 FSs $\{X_i\}_{i=1}^{n}$ are in descending order if $X_i \geq X_j$ for $\forall i < j$ by a ranking method. ∎

Any T1 FS ranking method can be used to find σ. In this chapter, Yager's first method [29] is used, which first computes the centroid of each T1 FS and then rank them to obtain the order of the corresponding T1 FSs.

To compute Y_{OFWA}, we first sort X_i in descending order and call them by the same name, but now $X_1 \geq X_2 \geq \cdots \geq X_n$ (W_i are not changed during this step); then, the FWA algorithm introduced in Sect. 2.2 can be used to compute Y_{OFWA}.

Example 5 For the same T1 FSs X_i and W_i used in Example 2, the OFWA Y_{OFWA} is shown as the dashed curve in Fig. 4c, which is different from Y_{FWA} [solid curve in Fig. 4c]. ∎

3.3 The Ordered Linguistic Weighted Average (OLWA)

As its name suggests, the OLWA is a combination of the OWA and the LWA.

Definition 10 An OLWA is defined as

$$\tilde{Y}_{OLWA} = \frac{\sum_{i=1}^{n} \tilde{W}_i \tilde{X}_{\sigma(i)}}{\sum_{i=1}^{n} \tilde{W}_{\sigma(i)}} \tag{25}$$

where $\sigma : \{1, \ldots, n\} \to \{1, \ldots, n\}$ is a permutation function such that $\{\tilde{X}_{\sigma(1)}, \tilde{X}_{\sigma(2)}, \ldots, \tilde{X}_{\sigma(n)}\}$ are in descending order. ∎

Definition 11 A group of IT2 FSs $\{\tilde{X}_i\}_{i=1}^{n}$ are in descending order if $\tilde{X}_i \geq \tilde{X}_j$ for $\forall i < j$ by a ranking method. ∎

Any IT2 FS ranking method can be used to find σ. In this chapter, the centroid-based ranking method [30] is used, which first computes the center of centroid of each IT2 FS and then ranks them to obtain the order of the corresponding IT2 FSs.

To compute the OLWA, we first sort all $\tilde{X}_i$ in descending order and call them by the same name, but now $\tilde{X}_1 \geq \tilde{X}_2 \geq \cdots \geq \tilde{X}_n$ (note that $\tilde{W}_i$ are not changed during this step); then, the LWA algorithm introduced in Sect. 2.3 can be used to compute the OLWA.

Example 6 For the same IT2 FSs $\tilde{X}_i$ and $\tilde{W}_i$ used in Example 3, the OLWA $\tilde{Y}_{OLWA}$ is shown as the dashed curve in Fig. 8c, which is different from $\tilde{Y}_{LWA}$ [solid curve in Fig. 8c]. ∎

4 Other Fuzzy Extensions of the OWA

There has been many works on fuzzy extensions of the OWA, e.g., linguistic ordered weighted averaging [31–34], uncertain linguistic ordered weighted averaging [35], and fuzzy linguistic ordered weighted averaging [36]; however, for these extensions, only the sub-criteria are modeled as T1 FSs whereas the weights are still crisp numbers. To the authors' best knowledge, Zhou et al. [14–16] are the first to consider fuzzy weights. Their approaches are introduced in this section for comparison purposes.

4.1 T1 Fuzzy OWAs

Zhou et al. [15, 16, 37] defined a T1 fuzzy OWA (T1FOWA) as:

Definition 12 Given T1 FSs $\{W_i\}_{i=1}^n$ and $\{X_i\}_{i=1}^n$, the membership function of a T1FOWA is computed by:

$$\mu_{Y_{T1FOWA}}(y) = \sup_{\frac{\sum_{i=1}^n w_i x_{\sigma(i)}}{\sum_{i=1}^n w_i} = y} \min(\mu_{W_1}(w_1), \ldots, \mu_{W_n}(w_n), \mu_{X_1}(x_1), \ldots, \mu_{X_n}(x_n)) \quad (26)$$

where $\sigma : \{1, \ldots, n\} \rightarrow \{1, \ldots, n\}$ is a permutation function such that $\{x_{\sigma(1)}, x_{\sigma(2)}, \ldots, x_{\sigma(n)}\}$ are in descending order. ∎

$\mu_{Y_{T1FOWA}}(y)$ can be understood from the Extension Principle [38], i.e., first all combinations of w_i and x_i whose OWA is y are found, and for the jth combination, the resulting y_j has a membership grade $\mu(y_j)$ which is the minimum of the corresponding $\mu_{X_i}(x_i)$ and $\mu_{W_i}(w_i)$. Then, $\mu_{Y_{T1FOWA}}(y)$ is the maximum of all these $\mu(y_j)$.

Y_{T1FOWA} can be computed efficiently using α-cuts [14], similar to the way they are used in computing the FWA. Denote $Y_{T1FOWA}(\alpha) = [y'_L(\alpha), y'_R(\alpha)]$ and use the same notations for α-cuts on X_i and W_i as in Fig. 2. Then,

$$y'_L(\alpha) = \min_{\forall w_i(\alpha) \in [c_i(\alpha), d_i(\alpha)]} \frac{\sum_{i=1}^n a_{\sigma(i)}(\alpha) w_i(\alpha)}{\sum_{i=1}^n w_i(\alpha)} \tag{27}$$

$$y'_R(\alpha) = \max_{\forall w_i(\alpha) \in [c_i(\alpha), d_i(\alpha)]} \frac{\sum_{i=1}^n b_{\sigma(i)}(\alpha) w_i(\alpha)}{\sum_{i=1}^n w_i(\alpha)} \tag{28}$$

$y'_L(\alpha)$ and $y'_R(\alpha)$ can also be computed using KM or EKM algorithms. Generally σ is different for different α in (27) and (28), because for each α the $a_i(\alpha)$ or $b_i(\alpha)$ are ranked separately.

Generally the OFWA and the T1FOWA give different outputs, as indicated by the following:

Theorem 1 *The OFWA and the T1FOWA have different results when at least one of the following two conditions occurs:*

1. *The left leg of X_i intersects the left leg of X_j, $i \neq j$.*
2. *The right leg of X_i intersects the right leg of X_j, $i \neq j$.* ∎

Proof: Because the proof for Condition 2 is very similar to that for Condition 1, only the proof for Condition 1 is given here.

Assume the left leg of X_i intersects the left leg of X_j at $\alpha = \lambda \in (0, 1)$, as shown in Fig. 9. Then, $a_i(\alpha) > a_j(\alpha)$ when $\alpha \in [0, \lambda)$ and $a_i(\alpha) < a_j(\alpha)$ when $\alpha \in (\lambda, 1]$.

For an $\alpha_1 \in [0, \lambda)$, $y'_L(\alpha_1)$ in (27) is computed as

$$y'_L(\alpha_1) = \min_{\forall w_i(\alpha_1) \in [c_i(\alpha_1), d_i(\alpha_1)]} \frac{\sum_{i=1}^n a_{\sigma_1(i)}(\alpha_1) w_i(\alpha_1)}{\sum_{i=1}^n w_i(\alpha_1)} \tag{29}$$

where $\sigma_1 : \{1, \dots, n\} \to \{1, \dots, n\}$ is a permutation function such that $\{a_{\sigma_1(1)}(\alpha_1), x_{\sigma_1(2)}(\alpha_1), \dots, x_{\sigma_1(n)}(\alpha_1)\}$ are in descending order. Because $a_i(\alpha_1) > a_j(\alpha_1)$, it follows that $\sigma_1(i) < \sigma_1(j)$.

For an $\alpha_2 \in (\lambda, 1]$, $y'_L(\alpha)$ in (27) is computed as

$$y'_L(\alpha_2) = \min_{\forall w_i(\alpha_2) \in [c_i(\alpha_2), d_i(\alpha_2)]} \frac{\sum_{i=1}^n a_{\sigma_2(i)}(\alpha_2) w_i(\alpha_2)}{\sum_{i=1}^n w_i(\alpha_2)} \tag{30}$$

where $\sigma_2 : \{1, \dots, n\} \to \{1, \dots, n\}$ is a permutation function such that $\{a_{\sigma_2(1)}(\alpha_2), a_{\sigma_2(2)}(\alpha_2), \dots, a_{\sigma_2(n)}(\alpha_2)\}$ are in descending order. Because $a_i(\alpha_2) < a_j(\alpha_2)$, it follows that $\sigma_2(i) > \sigma_2(j)$, i.e., $\sigma_1 \neq \sigma_2$.

Fig. 9 Illustration of intersecting X_i and X_j

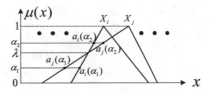

On the other hand, for Y_{OFWA}, no matter which ranking method is used, the permutation function σ is the same for all $\alpha \in [0, 1]$. Without loss of generality, assume $X_j \geq X_i$ by a ranking method. Then, in (24) $\sigma(i) > \sigma(j)$, and, for any $\alpha \in [0, 1]$, $y_L(\alpha)$ is computed as

$$y_L(\alpha) = \min_{\forall w_i(\alpha) \in [c_i(\alpha), d_i(\alpha)]} \frac{\sum_{i=1}^{n} a_{\sigma(i)}(\alpha) w_i(\alpha)}{\sum_{i=1}^{n} w_i(\alpha)} \tag{31}$$

Clearly, for any $\alpha \in [0, \lambda)$, $y_L(\alpha) \neq y'_L(\alpha)$ because $\sigma \neq \sigma_1$. Consequently, the left legs of Y_{OFWA} and Y_{T1FOWA} are different. ∎

The following example illustrates Theorem 1.

Example 7 X_i and W_i shown in Figs. 4a and b are used in this example to illustrate the difference between Y_{T1FOWA} and Y_{OFWA}. Y_{T1FOWA} is shown as the dotted curve in Fig. 4c. Note that it is quite different from Y_{OFWA} [dashed curve in Fig. 4c]. The difference is caused by the fact that the legs of X_3 cross the legs of X_1, X_2 and X_4, which causes the permutation function σ to change as α increases. ∎

Finally, observe two important points from Theorem 1:

1. Only the intersection of a left leg with another left leg, or a right leg with another right leg, would definitely lead to different Y_{T1FOWA} and Y_{OFWA}. The intersection of a left leg with a right leg does not lead to different Y_{T1FOWA} and Y_{OFWA}, as illustrated by Example 8.
2. Only the intersections of X_i may lead to different Y_{T1FOWA} and Y_{OFWA}. The intersections of W_i have no effect on this because the permutation function σ does not depend on W_i.

Example 8 Consider X_i shown in Fig. 10a and W_i shown in Fig. 10b. Y_{FWA} is shown as the solid curve in Fig. 10c, Y_{OFWA} the dashed curve, and Y_{T1FOWA} the dotted curve (the latter two are covered by the solid curve). Though X_i have some intersections, Y_{T1FOWA} is the same as Y_{OFWA} because no left (right) legs of X_i intersect. ∎

4.2 IT2 Fuzzy OWAs

Zhou et al. [16] defined the IT2 fuzzy OWA (IT2FOWA) as:

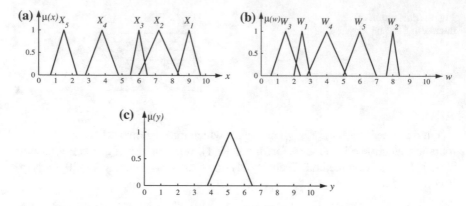

Fig. 10 Example 8, where Y_{FWA}, Y_{OFWA} and Y_{T1FOWA} give the same result: **a** X_i, **b** W_i, and **c** Y_{FWA} (solid curve), Y_{OFWA} (dashed curve) and Y_{T1FOWA} (dotted curve)

Definition 13 Given IT2 FSs $\{\tilde{W}_i\}_{i=1}^n$ and $\{\tilde{X}_i\}_{i=1}^n$, the membership function of an IT2FOWA is computed by:

$$\mu_{\tilde{Y}_{IT2FOWA}}(y) = \bigcup_{\forall W_i^e, X_i^e} \left[\sup_{\frac{\sum_{i=1}^n w_i x_{\sigma(i)}}{\sum_{i=1}^n w_i} = y} \min(\mu_{W_1^e}(w_1), \ldots, \mu_{W_n^e}(w_n), \mu_{X_1^e}(x_1), \ldots, \mu_{X_n^e}(x_n)) \right] \quad (32)$$

where W_i^e and X_i^e are embedded T1 FSs of $\tilde{W}_i$ and $\tilde{X}_i$, respectively, and $\sigma : \{1, \ldots, n\} \to \{1, \ldots, n\}$ is a permutation function such that $\{x_{\sigma(1)}, x_{\sigma(2)}, \ldots, x_{\sigma(n)}\}$ are in descending order. ∎

Comparing (32) with (26), observe that the bracketed term in (32) is a T1FOWA, and the IT2FOWA is the union of all possible T1FOWAs computed from the embedded T1 FSs of $\tilde{X}_i$ and $\tilde{W}_i$. The Wavy Slice Representation Theorem [39] for IT2 FSs is used implicitly in this definition.

$\tilde{Y}_{IT2FOWA}$ can be computed efficiently using α-cuts, similar to the way they were used in computing the LWA. Denote the α-cut on the UMF of $\tilde{Y}_{IT2FOWA}$ as $\overline{Y}_{OWA}(\alpha) = [y'_{Ll}(\alpha), y'_{Rr}(\alpha)]$ for $\forall \alpha \in [0, 1]$, the α-cut on the LMF of $\tilde{Y}_{IT2FOWA}$ as $\underline{Y}_{OWA}(\alpha) = [y'_{Lr}(\alpha), y'_{Rl}(\alpha)]$ for $\forall \alpha \in [0, h_{\min}]$, where $h_{\min}$ is defined in (16). Using the same notations for α-cuts on $\tilde{X}_i$ and $\tilde{W}_i$ as in Fig. 8, it is easy to show that

$$y'_{LI}(\alpha) = \min_{\forall w_i(\alpha) \in [c_{il}(\alpha), d_{ir}(\alpha)]} \frac{\sum_{i=1}^{n} a_{\sigma(i),l}(\alpha) w_i(\alpha)}{\sum_{i=1}^{n} w_i(\alpha)}, \alpha \in [0, 1] \qquad (33)$$

$$y'_{Rr}(\alpha) = \max_{\forall w_i(\alpha) \in [c_{il}(\alpha), d_{ir}(\alpha)]} \frac{\sum_{i=1}^{n} b_{\sigma(i),r}(\alpha) w_i(\alpha)}{\sum_{i=1}^{n} w_i(\alpha)}, \alpha \in [0, 1] \qquad (34)$$

$$y'_{Lr}(\alpha) = \min_{\forall w_i(\alpha) \in [c_{ir}(\alpha), d_{il}(\alpha)]} \frac{\sum_{i=1}^{n} a_{\sigma(i),r}(\alpha) w_i(\alpha)}{\sum_{i=1}^{n} w_i(\alpha)}, \alpha \in [0, h_{\min}] \qquad (35)$$

$$y'_{Rl}(\alpha) = \max_{\forall w_i(\alpha) \in [c_{ir}(\alpha), d_{il}(\alpha)]} \frac{\sum_{i=1}^{n} b_{\sigma(i),l}(\alpha) w_i(\alpha)}{\sum_{i=1}^{n} w_i(\alpha)}, \alpha \in [0, h_{\min}] \qquad (36)$$

$y'_{LI}(\alpha)$, $y'_{Rr}(\alpha)$, $y'_{Lr}(\alpha)$ and $y'_{Rl}(\alpha)$ can also be computed using KM or EKM algorithms. Because $\tilde{Y}_{IT2FOWA}$ computes the permutation function σ for each α separately, generally σ is different for different α.

Generally the OLWA and the IT2FOWA give different outputs, as indicated by the following:

Theorem 2 *The OLWA and the IT2FOWA have different results when at least one of the following four conditions occur:*

1. *The left leg of $\overline{X}_i$ intersects the left leg of $\overline{X}_j$, $i \neq j$.*
2. *The left leg of $\underline{X}_i$ intersects the left leg of $\underline{X}_j$, $i \neq j$.*
3. *The right leg of $\overline{X}_i$ intersects the right leg of $\overline{X}_j$, $i \neq j$.*
4. *The right leg of $\underline{X}_i$ intersects the right leg of $\underline{X}_j$, $i \neq j$.*   ∎

The correctness of Theorem 2 can be easily seen from Theorem 1, i.e., Condition 1 leads to different $y_{LI}(\alpha)$ and $y'_{LI}(\alpha)$ for certain α, Condition 2 leads to different $y_{Lr}(\alpha)$ and $y'_{Lr}(\alpha)$ for certain α, Condition 3 leads to different $y_{Rr}(\alpha)$ and $y'_{Rr}(\alpha)$ for certain α, and Condition 4 leads to different $y_{Rl}(\alpha)$ and $y'_{Rl}(\alpha)$ for certain α. Example 9 illustrates Theorem 2.

Example 9 $\tilde{X}_i$ and $\tilde{W}_i$ shown in Figs. 8a and b are used in this example to illustrate the difference between $\tilde{Y}_{OLWA}$ and $\tilde{Y}_{IT2FOWA}$. $\tilde{Y}_{IT2FOWA}$ is shown as the dotted curve in Fig. 8c. Note that it is quite different from $\tilde{Y}_{OLWA}$ [dashed curve in Fig. 8c]. The difference is caused by the fact that the legs of $\tilde{X}_3$ cross the legs of $\tilde{X}_1$, $\tilde{X}_2$ and $\tilde{X}_4$, since the permutation function σ changes as α increases.   ∎

Finally, observe also two important points from Theorem 2:

1. Only the intersection of a left leg with another left leg, or a right leg with another right leg, would definitely lead to different $\tilde{Y}_{IT2FOWA}$ and $\tilde{Y}_{OLWA}$. The intersection of a left leg with a right leg may not lead to different $\tilde{Y}_{IT2FOWA}$ and $\tilde{Y}_{OLWA}$, as illustrated by Example 10.
2. Only the intersections of $\tilde{X}_i$ may lead to different $\tilde{Y}_{IT2FOWA}$ and $\tilde{Y}_{OLWA}$. The intersections of $\tilde{W}_i$ have no effect on this because the permutation function σ does not depend on $\tilde{W}_i$.

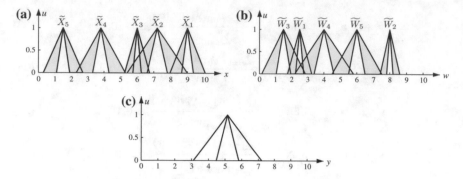

Fig. 11 Example 10, where IT2FOWA and OLWA give the same result: **a** $\tilde{X}_i$, **b** $\tilde{W}_i$, and **c** $\tilde{Y}_{LWA}$ (solid curve), $\tilde{Y}_{OLWA}$ (dashed curve) and $\tilde{Y}_{IT2FOWA}$ (dotted curve)

Example 10 Consider $\tilde{X}_i$ shown in Fig. 11a and $\tilde{W}_i$ shown in Fig. 11b. $\tilde{Y}_{LWA}$ is shown as the solid curve in Fig. 11c, $\tilde{Y}_{OLWA}$ the dashed curve, and $\tilde{Y}_{IT2FOWA}$ the dotted curve (the latter two are covered by the solid curve). Though $\tilde{X}_i$ have some intersections, $\tilde{Y}_{IT2FOWA}$ is the same as $\tilde{Y}_{OLWA}$. ∎

Example 11 In this final example, we compare the results of LWA, OLWA and IT2FOWA when

$$\{\tilde{X}_i\}|_{i=1,\ldots,4} \rightarrow \{\text{Tiny, Maximum amount, Fair amount, Medium}\}$$
$$\{\tilde{W}_i\}|_{i=1,\ldots,4} \rightarrow \{\text{Small, Very little, Sizeable, Huge amount}\}.$$

where the word FOUs are depicted in Fig. 12a and b. They are extracted from the 32-word vocabulary in [2, 3, 40], which is constructed from actual survey data. The corresponding $\tilde{Y}_{LWA}$ is shown in Fig. 12c as the solid curve, $\tilde{Y}_{OLWA}$ the dashed curve, and $\tilde{Y}_{IT2FOWA}$ the dotted curve. Observe that they are different from each other. ∎

4.3 Discussions

The T1 and IT2 fuzzy OWAs have been derived by considering each α-cut separately, whereas the OFWA and OLWA have been derived by considering each sub-criterion as a whole. Generally the two approaches give different results. Then, a natural question is: which approach should be used in practice?

We believe that it is more intuitive to consider an FS in its entirety during ranking of FSs. To the best of our knowledge, all ranking methods based on α-cuts deduce a single number to represent each FS and then sort these numbers to obtain the ranks of the FSs (see the Appendix). Each of these numbers is computed based only on

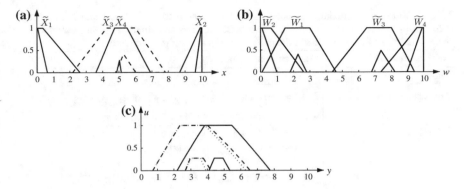

Fig. 12 Example 11: **a** $\tilde{X}_i$, **b** $\tilde{W}_i$, and, **c** $\tilde{Y}_{LWA}$ (solid curve), $\tilde{Y}_{OLWA}$ (dashed curve) and $\tilde{Y}_{IT2FOWA}$ (dotted curve)

the FS under consideration, i.e., no α-cuts on other FSs to be ranked are considered. Because in OFWA and OLWA the FSs are first ranked and then the WAs are computed, they coincide with our "FS in its entirety" intuition, and hence they are preferred in this chapter. Interestingly, this "FS in its entirety" intuition was also used implicitly in developing the linguistic ordered weighted averaging [32], the uncertain linguistic ordered weighted averaging [35], and the fuzzy linguistic ordered weighted averaging [36].

5 Conclusions

In this chapter, ordered novel weighted averages, including ordered interval weighted average, ordered fuzzy weighted average and ordered linguistic weighted average, as wells as procedures for computing them, have been introduced. They were compared with novel weighted averages and Zhou et al's fuzzy extensions of the OWA. Examples showed that our ONWAs may give different results from Zhou et al's extensions when the legs of the FSs have intersections. Because our extensions coincide with the "FS in its entirety" intuition, they are the suggested ones to use.

Ranking Methods for T1 FSs

Wang and Kerre [41–44] performed a very comprehensive study on ranking methods for T1 FSs. They partitioned over 35 ranking methods for T1 FSs into three classes:

1. *Class 1*: Reference set(s) is (are) set up, and each T1 FS is mapped into a crisp number based on the reference(s). The T1 FSs are then ranked according to the corresponding crisp numbers.

Table 1 Summary of ranking methods for T1 FSs. Note that for Classes 1 and 2, each T1 FS is first mapped into a crisp number, and then these numbers are sorted to obtain the ranks of the corresponding T1 FSs. For Class 3, the pairwise ranks are computed directly

Ranking method	Equation used for ranking
Class 1	
Jain's method [45, 46]	$f_J(A_i) = \sup_x \min(\mu_{A_{\max,J}}(x), \mu_{A_i}(x))$, where $\mu_{A_{\max,J}}(x) = \left(\frac{x}{x_{\max}}\right)^k$, in which $k > 0$ and $x_{\max}$ is the right end of the x domain
Chen's method [47]	$f_C(A_i) = [R(A_i) + 1 - L(A_i)]/2$, where $R(A_i) = \sup_x \min(\mu_{A_{\max,C}}(x), \mu_{A_i}(x))$, $L(A_i) = \sup_x \min(\mu_{A_{\min,C}}(x), \mu_{A_i}(x))$, $\mu_{A_{\max,C}}(x) = \left(\frac{x-x_{\min}}{x_{\max}-x_{\min}}\right)^k$, $\mu_{A_{\min,C}}(x) = \left(\frac{x_{\max}-x}{x_{\max}-x_{\min}}\right)^k$, $k > 0$, and $x_{\min}$ is the left end of the x domain
Kim and Park's method [48]	$f_{KP}(A_i) = k h_{A_i \cap A_{\max,KP}} + (1-k)(1 - h_{A_i \cap A_{\min,KP}})$, where $k \in [0, 1]$, $h_{A_i \cap A_{\max,KP}}$ is the height of $A_i \cap A_{\max,KP}$, $\mu_{A_{\max,KP}}(x) = \frac{x-x_{\min}}{x_{\max}-x_{\min}}$, and $\mu_{A_{\min,KP}}(x) = \frac{x_{\max}-x}{x_{\max}-x_{\min}}$
Class 2	
Adamo's method [49]	Let $A_{i\alpha}$ be an α-cut of a T1 FS A_i $f_A(A_i) = r(A_{i\alpha})$, where $r(A_{i\alpha})$ is the right end of $A_{i\alpha}$, and α can be any user-chosen number in $(0, 1]$
Yager's first method [29]	$f_Y(A_i) = \dfrac{\int_0^1 x\mu_{A_i}(x)dx}{\int_0^1 \mu_{A_i}(x)dx}$, where the domain of x is constrained in $[0, 1]$
Yager's second method [50, 51]	$f_Y(A_i) = \int_0^{h_{A_i}} m(A_{i\alpha})d\alpha$, where h_{A_i} is the height of A_i, and $m(A_{i\alpha})$ is the center of $A_{i\alpha}$
Fortemps and Roubens' method [52]	$f_{FR}(A_i) = \frac{1}{h_{A_i}} \int_0^{h_{A_i}} [r(A_{i\alpha}) - l(A_{i\alpha})]d\alpha$, where $l(A_\alpha)$ is the left end of $A_{i\alpha}$
Class 3	
	$d_H(A_i, A_j) \equiv \int_X [\mu_{A_i}(x) - \mu_{A_j}(x)]dx$, A^l and A^u are T1 FSs defined as $\mu_{A^l}(x) \equiv \sup_{y \leq x} \mu_{A_i}(y)$, $\mu_{A^u}(x) \equiv \sup_{y \geq x} \mu_{A_i}(y)$ $\widetilde{\max}(A, B)$ and $\widetilde{\min}(A, B)$ are T1 FSs defined as $\mu_{\widetilde{\max}(A,B)}(x) = \sup_{x=u \vee v} [\mu_A(u) \vee \mu_B(v)]$ $\mu_{\widetilde{\min}(A,B)}(x) = \sup_{x=u \wedge v} [\mu_A(u) \wedge \mu_B(v)]$
Nakamura's method [53]	$r(A_i, A_j) = \dfrac{k d_H(A_i^l, \widetilde{\min}(A_i^l, A_j^l)) + (1-k)d_H(A_i^u, \widetilde{\min}(A_i^u, A_j^u))}{k d_H(A_i^l, A_j^l) + (1-k)d_H(A_i^u, A_j^u)}$
Kolodziejczyk's method [54]	$r(A_i, A_j) =$ $\dfrac{d_H(A_i^l, \widetilde{\min}(A_i^l, A_j^l)) + d_H(A_i^u, \widetilde{\min}(A_i^u, A_j^u)) + d_H(A_i \cap A_j, \emptyset)}{d_H(A_i^l, A_j^l) + d_H(A_i^u, A_j^u) + 2d_H(A_i \cap A_j, \emptyset)}$
Saade and Schwarzlander's method [55]	$r(A_i, A_j) = d_H(A_i^l, \widetilde{\max}(A_i^l, A_j^l)) + d_H(A_i^u, \widetilde{\max}(A_i^u, A_j^u))$

2. *Class 2*: A function $f(A_i)$ is used to map a T1 FS A_i to a crisp number, which can then be ranked. No reference set(s) is (are) used in the mapping.
3. *Class 3*: T1 FSs A_i ($i = 1, \dots, M$) are ranked through pairwise comparisons.

They then proposed seven reasonable properties that a ranking method should satisfy [41]. Some simple ranking methods, which are also the most reasonable ones according to the seven properties [41, 42], are summarized in Table 1.

References

1. J.M. Mendel, D. Wu, Computing with words for hierarchical and distributed decision making, in *Computational Intelligence in Complex Decision Systems*, D. Ruan, Ed. (Paris, France: Atlantis Press, 2010)
2. J.M. Mendel, D. Wu, *Perceptual Computing: Aiding People in Making Subjective Judgments* (Wiley-IEEE Press, Hoboken, NJ, 2010)
3. D. Wu, Intelligent systems for decision support, Ph.D. dissertation, University of Southern California, (Los Angeles, CA, May 2009)
4. D. Wu, J.M. Mendel, Aggregation using the linguistic weighted average and interval type-2 fuzzy sets. IEEE Trans. Fuzzy Syst. **15**(6), 1145–1161 (2007)
5. D. Wu, J.M. Mendel, Corrections to aggregation using the linguistic weighted average and interval type-2 fuzzy sets. IEEE Trans. Fuzzy Syst. **16**(6), 1664–1666 (2008)
6. D. Wu, J.M. Mendel, Computing with words for hierarchical decision making applied to evaluating a weapon system. IEEE Trans. Fuzzy Syst. **18**(3), 441–460 (2010)
7. D. Wu, J.M. Mendel, Social judgment advisor: an application of the perceptual computer, in Proceedings of IEEE World Congress on Computational Intelligence, (Barcelona, Spain, July 2010)
8. D. Filev, R. Yager, On the issue of obtaining OWA operator weights. Fuzzy Sets Syst. **94**, 157–169 (1998)
9. X. Liu, The solution equivalence of minimax disparity and minimum variance problems for OWA operators. Int. J. Approx. Reason. **45**, 68–81 (2007)
10. P. Majlender, OWA operators with maximal Renya entropy. Fuzzy Sets Syst. **155**, 340–360 (2005)
11. V. Torra, Y. Narukawa, *Modeling Decisions: Information Fusion and Aggregation Operators* (Springer, Berlin, 2007)
12. R. Yager, On ordered weighted averaging aggregation operators in multi-criteria decision making. IEEE Trans. Syst. Man Cybernet. **18**, 183–190 (1988)
13. R. Yager, J. Kacprzyk, *The Ordered Weighted Averaging Operators: Theory and Applications* (Kluwer, Norwell, MA, 1997)
14. S.-M. Zhou, F. Chiclana, R. I. John, and J. M. Garibaldi, A practical approach to type-1 OWA operation for soft decision making, in *Proceedings of 8th International FLINS Conference on Computational Intelligence in Decision and Control*, (Madrid, Spain, 2008) pp. 507–512
15. S.-M. Zhou, F. Chiclana, R.I. John, J.M. Garibaldi, Type-1 OWA operators for aggregating uncertain information with uncertain weights induced by type-2 linguistic quantifiers. Fuzzy Sets Syst. **159**(24), 3281–3296 (2008)
16. S.-M. Zhou, F. Chiclana, R.I. John, J.M. Garibaldi, Type-2 OWA operators–aggregating type-2 fuzzy sets in soft decision making, in *Proceedings of IEEE International Conference on Fuzzy Systems* (Hong Kong, June, 2008), pp. 625–630
17. D. Dubois, H. Prade, A review of fuzzy sets aggregation connectives. Informat. Sci. **36**, 85–121 (1985)
18. N.N. Karnik, J.M. Mendel, Centroid of a type-2 fuzzy set. Informat. Sci. **132**, 195–220 (2001)

19. J.M. Mendel, *Uncertain Rule-Based Fuzzy Logic Systems: Introduction and New Directions* (Prentice-Hall, Upper Saddle River, NJ, 2001)
20. D. Wu, J.M. Mendel, *Enhanced Karnik-Mendel Algorithms for interval type-2 fuzzy sets and systems, in Proc* (CA, June, NAFIPS, San Diego, 2007), pp. 184–189
21. D. Wu, J.M. Mendel, Enhanced Karnik-Mendel algorithms. IEEE Trans. Fuzzy Syst. **17**(4), 923–934 (2009)
22. W.M. Dong, F.S. Wong, Fuzzy weighted averages and implementation of the extension priniple. Fuzzy Sets Syst. **21**, 183–199 (1987)
23. Y.-Y. Guh, C.-C. Hon, E.S. Lee, Fuzzy weighted average: the linear programming approach via Charnes and Cooper's rule. Fuzzy Sets Syst. **117**, 157–160 (2001)
24. Y.-Y. Guh, C.-C. Hon, K.-M. Wang, E.S. Lee, Fuzzy weighted average: a max-min paired elimination method. J. Comput. Math. Appl. **32**, 115–123 (1996)
25. D.H. Lee, M.H. Kim, Database summarization using fuzzy ISA hierarchies. IEEE Trans. Syst. Man Cybernet. B **27**, 68–78 (1997)
26. T.-S. Liou, M.-J.J. Wang, Fuzzy weighted average: an improved algorithm. Fuzzy Sets Syst. **49**, 307–315 (1992)
27. F. Liu, J.M. Mendel, Aggregation using the fuzzy weighted average, as computed using the Karnik-Mendel Algorithms. IEEE Trans. Fuzzy Syst. **12**(1), 1–12 (2008)
28. G.J. Klir, B. Yuan, *Fuzzy Sets and Fuzzy Logic: Theory and Applications* (Prentice-Hall, Upper Saddle River, NJ, 1995)
29. R. Yager, Ranking fuzzy subsets over the unit interval, in *Proceedings of IEEE Conference on Decision and Control*, vol. 17, 1978, pp. 1435–1437
30. D. Wu, J.M. Mendel, A comparative study of ranking methods, similarity measures and uncertainty measures for interval type-2 fuzzy sets. Informat. Sci. **179**(8), 1169–1192 (2009)
31. G. Bordogna, M. Fedrizzi, G. Pasi, A linguistic modelling of consensus in group decision making based on OWA operator. IEEE Trans. Syst. Man Cybern. **27**, 126–132 (1997)
32. F. Herrera, A sequential selection process in group decision making with linguistic assessment. Informat. Sci. **85**, 223–239 (1995)
33. F. Herrera, J. Verdegay, Linguistic assessments in group decision, in *Proceedings of 1st European Congress on Fuzzy and Intelligent Technologies*, Aachen, pp. 941–948 (1993)
34. F. Herrera, L. Martinez, A 2-tuple fuzzy linguistic representation model for computing with words. IEEE Trans. Fuzzy Syst. **8**(6), 746–752 (2000)
35. Z. Xu, Uncertain linguistic aggregation operators based approach to multiple attribute group decision making under uncertain linguistic environment. Informat. Sci. **168**, 171–184 (2004)
36. D. Ben-Arieh, Z. Chen, Linguistic-labels aggregation and consensus measure for autocratic decision making using group recommendations. IEEE Trans. Syst. Man Cybern. A **36**(3), 558–568 (2006)
37. R.I. John, S.-M. Zhou, J.M. Garibaldi, F. Chiclana, Automated group decision support systems under uncertainty: trends and future research. Int. J. Comput. Intell. Res. **4**(4), 357–371 (2008)
38. L.A. Zadeh, The concept of a linguistic variable and its application to approximate reasoning-1. Informat. Sci. **8**, 199–249 (1975)
39. J.M. Mendel, R.I. John, F. Liu, Interval type-2 fuzzy logic systems made simple. IEEE Trans. Fuzzy Syst. **14**(6), 808–821 (2006)
40. F. Liu, J.M. Mendel, Encoding words into interval type-2 fuzzy sets using an interval approach. IEEE Trans. Fuzzy Syst. **16**(6), 1503–1521 (2008)
41. X. Wang, E.E. Kerre, Reasonable properties for the ordering of fuzzy quantities (I). Fuzzy Sets Syst. **118**, 375–387 (2001)
42. X. Wang, E.E. Kerre, Reasonable properties for the ordering of fuzzy quantities (II). Fuzzy Sets Syst. **118**, 387–405 (2001)
43. X. Wang, A comparative study of the ranking methods for fuzzy quantities, Ph.D. dissertation, University of Gent, (1997)
44. X. Wang, E.E. Kerre, On the classification and the dependencies of the ordering methods, in *Fuzzy Logic Foundations and Industrial Applications*, ed. by D. Ruan (Kluwer Academic Publishers, Dordrech, 1996), pp. 73–88

45. R. Jain, Decision making in the presence of fuzzy variables. IEEE Trans. Syst. Man Cybernet. **6**, 698–703 (1976)
46. R. Jain, A procedure for multiple-aspect decision making using fuzzy set. Int. J. Syst. Sci. **8**, 1–7 (1977)
47. S. Chen, Ranking fuzzy numbers with maximizing set and minimizing set. Fuzzy Sets Syst. **17**, 113–129 (1985)
48. J. Kim, Y. Moon, B.P. Zeigler, Designing fuzzy net controllers using genetic algorithms. IEEE Control Syst. Mag. **15**(3), 66–72 (1995)
49. J. Adamo, Fuzzy decision trees. Fuzzy Sets Syst. **4**, 207–219 (1980)
50. R. Yager, On choosing between fuzzy subsets. Kybernetes **9**, 151–154 (1980)
51. R. Yager, A procedure for ordering fuzzy sets of the unit interval. Informat. Sci. **24**, 143–161 (1981)
52. P. Fortemps, M. Roubens, Ranking and defuzzification methods based on area compensation. Fuzzy Sets Syst. **82**, 319–330 (1996)
53. K. Nakamura, Preference relations on a set of fuzzy utilities as a basis for decision making. Fuzzy Sets Syst. **20**(2), 147–162 (1986)
54. W. Kolodziejczyk, Orlovsky's concept of decision-making with fuzzy preference relation–further results. Fuzzy Sets Syst. **19**(1), 11–20 (1986)
55. J. Saade, H. Schwarzlander, Ordering fuzzy sets over the real line: an approach based on decision making under uncertainty. Fuzzy Sets Syst. **50**(3), 237–246 (1992)

On the Comparison of Model-Based and Model-Free Controllers in Guidance, Navigation and Control of Agricultural Vehicles

Erkan Kayacan, Erdal Kayacan, I-Ming Chen, Herman Ramon and Wouter Saeys

Abstract In a typical agricultural field operation, an agricultural vehicle must be accurately navigated to achieve an optimal result by covering with minimal overlap during tillage, fertilizing and spraying. To this end, a small scale tractor-trailer system is equipped by using off the shelf sensors and actuators to design a fully autonomous agricultural vehicle. To alleviate the task of the operator and allow him to concentrate on the quality of work performed, various systems were developed for driver assistance and semi-autonomous control. Real-time experiments show that a controller, which gives a satisfactory trajectory tracking performance for a straight line, gives a large steady-state error for a curved line trajectory. On the other hand, if the controller is aggressively tuned to decrease the tracking error for the curved lines, the controller gives oscillatory response for the straight lines. Although existing autonomous agricultural vehicles use conventional controllers, learning control algorithms are required to handle different trajectory types, environmental uncertainties, such as variable crop and soil conditions. Therefore, adaptability is a must rather than a choice in agricultural operations. In terms of complex mechatronics systems, e.g. an agricultural tractor-trailer system, the performance of model-based

E. Kayacan
Coordinated Science Lab, Distributed Autonomous Systems Lab,
University of Illinois at Urbana -Champaign, Urbana, IL 61801, USA
e-mail: erkank@illinois.edu

E. Kayacan (✉) · I.-M. Chen
School of Mechanical & Aerospace Engineering, Nanyang Technological University,
Singapore 639798, Singapore
e-mail: erdal@ntu.edu.sg

I.-M. Chen
e-mail: michen@ntu.edu.sg

H. Ramon · W. Saeys
Department of Biosystems, Division of Mechatronics, Biostatistics and Sensors,
KU Leuven, 3001 Leuven, Belgium
e-mail: herman.ramon@kuleuven.be

W. Saeys
e-mail: wouter.saeys@kuleuven.be

© Springer International Publishing AG 2018
R. John et al. (eds.), *Type-2 Fuzzy Logic and Systems*,
Studies in Fuzziness and Soft Computing 362,
https://doi.org/10.1007/978-3-319-72892-6_3

and model-free control, i.e. nonlinear model predictive control and type-2 neuro-fuzzy control, is compared and contrasted, and eventually some design guidelines are also suggested.

1 Introduction

Agriculture is the oldest, and also still the most important, economic activity of the modern humankind society. Archaeological excavations show that we, as humankind, started this thrilling adventure approximately 11,500 years ago. This decision was not only to stop being hunters and gatherers but also to start adapting the nature to our needs instead of only adapting ourselves to the facts of the wild life. This adventure started with wild barley, wheat and lentils in the South Asia (Fertile crescent and Chogha Golan) [1–3]. Our new skill, the skill of dealing with the soil and growing domestic plants instead of eating only wild ones, was the first step of our civilization which caused a domino effect such as paving the way for living as clans in villages and even the rise of complex religions.

Whereas average life expectancy was around 25 years in the Paleolithic and Neolithic eras, thanks to modern medicine, in particular Alexander Fleming who discovered penicillin, it has reached to 80 years in the last century [4]. In other words, our world is constantly being overcrowded. According to the United Nations Food and Agriculture Organization (FAO), our world has to double food production by 2050 to meet rising demand. Since it is an obvious fact that we can no longer clear more forest, one of the possible solutions is to increase the overall agricultural production efficiency among which the application of intelligent agricultural vehicles.

Considering the high demand for increased efficiency, productivity and safety in farming operations, a precise trajectory tracking is needed for agricultural vehicles to improve quality meanwhile reducing cost. When the motivations are carefully examined, the following requirements can be identified for an autonomous production machine, such as a tractor-trailer system: smart (intelligent) and productive (automation). In light of these aforementioned conditions, theoretical and practical control and design methods, i.e. model-free and model-based methods, are proposed throughout this article.

1.1 Role of Robots in Agriculture

Agriculture is not only a vital economic activity of a civilized society but also a necessity for our survival. Therefore, technological developments have always been playing an important role to make the most of our land even in challenging geographical locations. Our aim has always been to use our land in a more efficient way under significant climate and pre-assumed meteorological conditions.

The use of production machines and intelligent vehicles in agriculture is always promising as it allows us to make simultaneous operations that cannot be performed by a human operator. For instance, when working with an agricultural machine (*e.g.* combine harvester), apart from navigating the machine, the operator must also supervise the work performed by the machine. To be a skilled operator, even for a particular agricultural production machine, is not sufficient since the operator must always adapt the machine settings due to time-varying crop and soil characteristics as well as environmental conditions. Switching paying attention to between the steering and the machine control results in an increase in the deviation from the optimal path in practice. To alleviate the task of the operator and allow him to concentrate on the quality of work performed, provision of some autonomous functions to an agricultural vehicle is the main task of the robotic system. In this respect, a driver assistance and semi-autonomous control system for an agricultural robot will be developed in this article. To dispose a fully autonomous system, a tractor is equipped with off the shelf actuators and sensors to achieve the aforementioned goals. On behalf of an operator, the developed advanced learning control algorithms are implemented in real-time to deal with changing soil conditions as well as longitudinal speed. All the aforementioned challenges tell us the same thing: *adaptability is a must rather than a choice.*

1.2   Why Do We Need Agricultural Robots?

There are at least four reasons that ensure the necessity of using autonomous agricultural vehicles in the future:

1. Constantly rising energy and labor costs (need for efficient machines)
2. Continuously adapt the machine settings (multitasking)
3. Maintain the fixed performance and accuracy (a human operator may get bored or tired after some time especially under challenging working conditions, *e.g.* under hot and sunny conditions)
4. Not possible to increase the size of the machines (limited road capacity)

1.3   What Are the Requirements of the Agricultural Vehicles?

Without exception, all agricultural operations have a strict requirement: accurate navigation. For instance, throughout tillage, fertilizing and spraying, the production machine must be operated with a high accuracy to avoid overlapping field operations. The field rows must be nicely parallel and evenly distributed so that for example the weed rows can be easily driven between them. In fact, this requirement is challenging as it can be observed that there is always considerable overlap and variation in plant distances in the field even in manual operation. The reason is that these vehicles have to operate in hilly, bumpy and sometimes muddy off-road conditions as well as they generally have to deal with thew dynamics of a trailer.

1.4   Literature Review

The first harvesting robot was introduced in The United States of America to harvest citrus [5]. After this successful implementation, it was also used to harvest apples in France in 1985 [6]. 2009, a robotic arm, which is capable of harvesting asparagus, was developed by the Industrial Technology Center of Nagasaki in Japan [7]. Afterwards, in 2011, a prototype robotic platform, which has the ability to detect spherical fruits by benefiting from image processing, was developed in [8]. It is concluded that the proposed platform can increase the overall efficiency by reducing the spent time for harvesting. As a vision-based method, in another study, detection of red and bicoloured apples on tree with an RGB-D camera has been reported [9]. Furthermore, an agribot has been developed by Birla Institute of Technology and Science to minimize the labor of farmers and increase the accuracy of the work [10]. As can be seen from the previous implementations, there have existed significant research and development in agricultural robotics. One of the most challenging tasks is to guide the mobile robotic platforms accurately on different soil conditions.

The main goal of guidance of agricultural vehicles is to drive the vehicle on an agricultural field for specific purposes by keeping it as close as to the target trajectory. There are numerous implementations of multitasking path planning for multi-vehicle cases [11, 12]. In one of them, the path planning is carried out just for one vehicle, leading vehicle, the rest of the vehicles follow it by ensuring the desired relative distances. A master-slave navigation system has been proposed in [13] where the automated slave vehicle always follows the master vehicle whether the master vehicle is autonomous or not. Another stable controller for a four-wheel mobile robot to track between rows on a field has been designed in [14] while a nonlinear model predictive controller has been proposed for a tractor-trailer system in [15]. Moreover, online learning algorithms have been integrated into control algorithms. A fuzzy controller has been designed where its membership functions (MFs) have been adapted to changing working conditions [16]. However, this paper was lack of analyzing the robustness of the proposed learning algorithm considering different environment conditions. A guidance method based on a grid map of the agricultural field has been proposed in [12] in which the grid information is used to make a feasible path from the starting point of the vehicle to the desired destination in the field. Moreover, an autonomous orchard vehicle has been developed to help fruit production in which the perception system is based on global positioning system and a two-dimensional laser scanner [17]. It can be concluded from all previous studies that learning algorithms must be used to design a controller for the purpose of guidance regarding different working conditions to obtain accurate trajectory tracking performance. After giving the design details of the autonomous agricultural vehicle, whereas we will elaborate different control algorithms to accurately navigate the agricultural vehicle under certain uncertainties in the working environment, interested readers may refer to [18] for a detailed analysis about the role of global navigation satellite systems (GNSSs) in the navigation strategies of agricultural robots.

Amongst the two well-known inference methods, as a learning model-free controller, Takagi-Sugeno-Kang (TSK) fuzzy structure has significant advantages over its Mamdani counterpart as it has tunable weights on the consequent part of the rules which allows us to update them using appropriate optimization algorithms [19]. Consequently, they are preferred in real-time application where the working conditions vary over the operation. What is more, TSK models are computationally more efficient. Considering the recent advances and proved capabilities of type-2 fuzzy logic controllers (T2FLCs) over their type-1 counterparts [20–26], we prefer to use a TSK T2FLC to handle uncertainties in the autonomous tractor-trailer system in this paper. On the other hand, as a model-based approach, a nonlinear model predictive controller (NMPC) is preferred as an advanced control algorithm. Some parameters are estimated using a nonlinear moving horizon estimator (NMHE), and fed to the model which is being used by the NMPC. The overall scheme is a learning model-based controller.

Model-based and model-free control approaches are compared and contrasted for wet clutch control problem [27]. However, the parameter update strategy in the model-free approaches considered in [27], genetic-based machine learning and reinforcement learning, are different than the method used in this paper. For instance, whereas agents take actions in an environment to maximize a cumulative reward in reinforcement learning, Lyapunov stability-based learning rules are used in the type-2 fuzzy structure in this paper which are shown to be stable using a candidate Lyapunov function.

1.5 Motivation

In terms of complex mechatronics systems, the performance of model-based and model-free control, i.e. nonlinear model predictive control and type-2 neuro-fuzzy control, is compared and contrasted by means of their design and implementation simplicity and efficiency. Moreover, some design guidelines are also suggested for the control complex mechatronic systems where there exist more than one subsystem.

1.6 Organization of the Paper

Section 2 gives the system description of the tractor-trailer system. Section 3 explains the self learning model-free and model-based algorithms; some guidelines for a controller design selection are also suggested. Finally, some conclusions are drawn from this study in Sect. 4.

2 Prototyped Autonomous Agricultural Vehicle

In order to be used during tillage, fertilizing or spraying, a small scale tractor with a trailer shown in Fig. 1 is equipped with relatively cheap sensors resulting in a fully autonomous agricultural ground robotic system. The main expectation from the designed vehicle is to follow a predetermined trajectory in outdoor environment with a high accuracy to decrease overlap during agricultural operations.

2.1 Localization

Localization and positioning systems are broadly categorized into two groups: local and global. Whereas image processing, lazer, etc. belong to local positioning systems, global positioning systems make use of satellite systems. Thanks to the recent developments in the field of GNSSs we have up to cm accuracy in real-time kinematic (RTK) GNSSs to navigate our tractor-trailer system precisely.

The requirements are to model the system, identify its parameters and design learning controllers for the system shown in Fig. 1. As this tractor has hydraulic wheel and steering, and is four-wheel-drive, it is representative for many modern agricultural vehicles. The most suitable places for mounting GNSS antennas for the tractor and trailer are the tractor rear axle center and the trailer rear axle center, respectively. Since the horizontal accuracy for civilian GPS is still around 4 meter, we have decided to use RTK differential GPS (DGPS) in our system. The resulting accuracy is 0.03 m according to the specifications of the manufacturer. In order to receive the correction signals via internet, we have preferred Flepos network by using a *Digi Connect WAN 3G* modem.

As the real-time controller, PXI platform (National Instruments Corporation, Austin, TX, USA) is selected. The GNSS and the modem are connected to the real-time controller via serial connection. The main responsibility of the real-time controller is to receive and process all the necessary sensory data, such as steering angles, GNSS measurements, etc., and to generate the control signals for the tractor and

Fig. 1 The tractor-trailer system

trailer actuators separately. The control algorithms are implemented in *LabVIEW*^{TM} version 2011 (National Instruments, Austin, TX, USA). The working frequency of the overall control system is chosen as 5-Hz.

2.2 Steering Mechanisms

We have preferred to use a potentiometer, which is mounted on the front axle, to measure the tractor front wheel angle. An inductive sensor is used to measure the angle between the trailer and its drawbar. Both sensors have 1° precision. The rpm of the diesel engine has been measured by using a hall effect sensor (Hamlin, USA) which is connected to the shaft between the diesel engine and oil pump. Figure 2 shows the potentiometers and the hydrostat spindle actuator.

Low level controllers, proportional-integral (PI) controllers, generate the voltage for the electro-hydraulic valves based on the difference between the reference and measured steering angles. The longitudinal velocity of the tractor is measured by encoders mounted on the rear wheels of the tractor. A low level controller (PID) generates the voltage for the spindle actuator (LINAK A/S, Silkeborg, Denmark) taking into account the difference between the reference and measured pedal positions. The pedal position is measured by a magnetic sensor.

Fig. 2 Trailer actuator (top right), potentiometer (bottom left) and hydrostat spindle actuator (bottom right)

3 Self-learning Control Algorithms

The current commercial systems use simple controllers to minimize the deviation from the target path by adjusting the steering angle. These systems work well for the following straight lines under uniform soil conditions with a constant speed. However, when the soil conditions or speed change, the controllers must be tuned again. Furthermore, they use independent controllers for the absolute steering of the tractor and the relative steering of the trailer. Since both controllers will exhibit selfish behavior, this often leads to a sub-optimal result, especially for curved target paths in which the steering action of the tractor works against that of the trailer.

As a solution to the selfish behavior of the decentralized and static control algorithms, self-learning controllers have been designed in this investigation. A learning

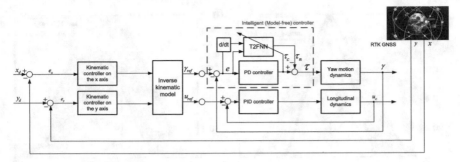

Fig. 3 Block diagram of the model-free controller

control algorithm, no matter it is model-based or model-free, is more than welcome as it will adapt itself against the parameter, crop and soil condition variations.

3.1 Model-Free Learning: Type-2 Fuzzy Neural Network Control

The proposed control scheme used in this part of the study is illustrated in Fig. 3. Since we have realized that the performance progress comes more from the yaw dynamics control accuracy of the overall control system, we have preferred to use only a conventional proportional-integral-derivative (PID) controller for the longitudinal dynamics, and design the intelligent model-free controller for the yaw dynamics. In the yaw dynamics control, a PD controller is used to guarantee the stability of the system during the initial learning. After a finite time, a type-2 fuzzy neural network (T2FNN) takes the control responsibility of the system, and the output of the PD controller goes to zero. Such a control scheme is called feedback error learning [28]. Thanks to the model-free structure of the controller, the dynamics and interactions between the subsystems are learnt online, and the optimal control signal is applied to the system. An outer loop for both the x and the y axes is also designed to correct the trajectory following errors on the relevant axes.

In the designed T2FLC, a triangular MF is preferred. There are two different approaches to construct type-2 triangular MFs. One is to blur the width of the MF Fig. 4a while the other is to blur the center of the MF Fig. 4a. In Fig. 4, the red line represents the upper MF, and the blue line shows the lower MF. Their corresponding membership values are $\overline{\mu}(x)$ and $\underline{\mu}(x)$, respectively.

The strength of the rule R_{ij} is calculated as a T-norm of the MFs in the premise part by using a multiplication operator:

$$\underline{W}_{ij} = \underline{\mu_{1i}(x_1)}\ \underline{\mu_{2j}(x_2)} \quad \text{and} \quad \overline{W}_{ij} = \overline{\mu_{1i}(x_1)}\ \overline{\mu_{2j}(x_2)} \tag{1}$$

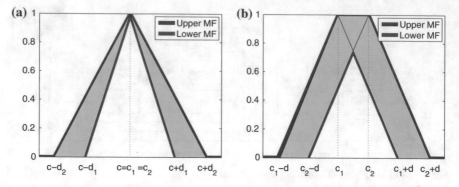

Fig. 4 A type-2 fuzzy triangular MF with uncertain width (**a**) and uncertain center (**b**)

The type-2 fuzzy triangular membership values $\underline{\mu_{1i}(x_1)}$, $\overline{\mu_{1i}(x_1)}$, $\underline{\mu_{2j}(x_2)}$, and $\overline{\mu_{2j}(x_2)}$ of the inputs x_1 and x_2 in the above expression have the following appearance:

$$\underline{\mu_{1i}(x_1)} = \begin{cases} 1 - \left| \frac{x_1 - c_{1i}}{\underline{d_{1i}}} \right| & |x_1 - c_{1i}| < \underline{d_{1i}} \\ 0 & otherwise \end{cases} \tag{2}$$

$$\overline{\mu_{1i}(x_1)} = \begin{cases} 1 - \left| \frac{x_1 - c_{1i}}{\overline{d_{1i}}} \right| & |x_1 - c_{1i}| < \overline{d_{1i}} \\ 0 & otherwise \end{cases}$$

$$\underline{\mu_{2j}(x_2)} = \begin{cases} 1 - \left| \frac{x_2 - c_{2j}}{\underline{d_{2j}}} \right| & |x_2 - c_{2j}| < \underline{d_{2j}} \\ 0 & otherwise \end{cases}$$

$$\overline{\mu_{2j}(x_2)} = \begin{cases} 1 - \left| \frac{x_2 - c_{2j}}{\overline{d_{2j}}} \right| & |x_2 - c_{2j}| < \overline{d_{2j}} \\ 0 & otherwise \end{cases}$$

Since we do not prefer to use an iterative type-reduction method in this paper, we prefer to use an approximated model of a type-2 fuzzy logic system which is denoted as A2-C0 fuzzy system. The rationale is to be able to use an optimization algorithm, which is a sliding mode control theory-based one in this paper, to tune the antecedent and consequent parameters. The fuzzy *If-Then* rule is defined as follows:

$$R_{ij} : \quad \text{If } x_1 \text{ is } \tilde{A}_{1i} \text{ and } x_2 \text{ is } \tilde{A}_{2j}, \quad \text{then } f_{ij} = d_{ij} \tag{3}$$

The output of the network is calculated as follows:

$$\tau_n = \int_{W_{11} \in [\underline{W}_{11}, \overline{W}_{11}]} \cdots \int_{W_{IJ} \in [\underline{W}_{IJ}, \overline{W}_{IJ}]} 1 \bigg/ \frac{\sum_{i=1}^{I} \sum_{j=1}^{J} W_{ij}(x) f_{ij}}{\sum_{i=1}^{I} \sum_{j=1}^{J} W_{ij}(x)} \tag{4}$$

where f_{ij} is given by the *If-Then* rule. The inference engine used in this paper replaces the type-reduction which is given as:

$$\tau_n = \frac{q(t) \sum_{i=1}^{I} \sum_{j=1}^{J} \underline{W}_{ij} f_{ij}}{\sum_{i=1}^{I} \sum_{j=1}^{J} \underline{W}_{ij}} + \frac{(1 - q(t)) \sum_{i=1}^{I} \sum_{j=1}^{J} \overline{W}_{ij} f_{ij}}{\sum_{i=1}^{I} \sum_{j=1}^{J} \overline{W}_{ij}} \tag{5}$$

The design parameter q, weights the sharing of the lower and the upper firing levels of each fired rule. After the normalization of (5), the output signal of the T2FNN will obtain the following form:

$$\tau_n = q(t) \sum_{i=1}^{I} \sum_{j=1}^{J} f_{ij} \widetilde{\underline{W}}_{ij} + (1 - q(t)) \sum_{i=1}^{I} \sum_{j=1}^{J} f_{ij} \widetilde{\overline{W}}_{ij} \tag{6}$$

where $\widetilde{\underline{W}}_{ij}$ and $\widetilde{\overline{W}}_{ij}$ are the normalized values of the lower and the upper output signals of the neuron ij::

$$\widetilde{\underline{W}}_{ij} = \frac{\underline{W}_{ij}}{\sum_{i=1}^{I} \sum_{j=1}^{J} \underline{W}_{ij}} \quad \text{and} \quad \widetilde{\overline{W}}_{ij} = \frac{\overline{W}_{ij}}{\sum_{i=1}^{I} \sum_{j=1}^{J} \overline{W}_{ij}}$$

The following vectors can be specified:

$$\widetilde{\underline{W}}(t) = \left[\widetilde{\underline{W}}_{11}(t) \ \widetilde{\underline{W}}_{12}(t) \ldots \ \widetilde{\underline{W}}_{21}(t) \ \ldots \ \widetilde{\underline{W}}_{ij}(t) \ \ldots \ \widetilde{\underline{W}}_{IJ}(t) \right]^T$$

$$\widetilde{\overline{W}}(t) = \left[\widetilde{\overline{W}}_{11}(t) \ \widetilde{\overline{W}}_{12}(t) \ldots \ \widetilde{\overline{W}}_{21}(t) \ \ldots \ \widetilde{\overline{W}}_{ij}(t) \ \ldots \ \widetilde{\overline{W}}_{IJ}(t) \right]^T$$

$$F = [f_{11} f_{12} \ \cdots \ f_{21} \ \cdots \ f_{ij} \ \cdots \ f_{IJ}]$$

The following assumptions have been used in this investigation: Both the input signals $x_1(t)$ and $x_2(t)$, and their time derivatives can be considered bounded:

$$|x_1(t)| \leq \widetilde{B}_x, \quad |x_2(t)| \leq \widetilde{B}_x \quad \forall t \tag{7}$$

$$|\dot{x}_1(t)| \leq \widetilde{B}_{\dot{x}}, \quad |\dot{x}_2(t)| \leq \widetilde{B}_{\dot{x}} \quad \forall t \tag{8}$$

where $\tilde{B}_x$ and $\tilde{B}_{\dot{x}}$ are assumed to be some known positive constants. It is obvious that $0 < \widetilde{W}_{ij} \leq 1$ and $0 < \overline{\widetilde{W}}_{ij} \leq 1$. In addition, it can be easily seen that $\sum_{i=1}^{I} \sum_{j=1}^{J} \underline{\widetilde{W}}_{ij} = 1$ and $\sum_{i=1}^{I} \sum_{j=1}^{J} \overline{\widetilde{W}}_{ij} = 1$. It is also considered that, τ and $\dot{\tau}$ will be bounded signals too, i.e.

$$|\tau(t)| < B_\tau, \quad |\dot{\tau}(t)| < B_{\dot{\tau}} \quad \forall t \tag{9}$$

where B_τ and $B_{\dot{\tau}}$ are some known positive constants.

3.1.1 Proposed Sliding Mode Control (SMC) Theory-Based Learning Algorithm

The zero value of the learning error coordinate $\tau_c(t)$ can be defined as a time-varying sliding surface, i.e.,

$$S_c(\tau_n, \tau) = \tau_c(t) = \tau_n(t) + \tau(t) = 0 \tag{10}$$

The sliding surface is defined as follows:

$$S_p(e, \dot{e}) = \dot{e} + \chi e \tag{11}$$

where χ is a positive constant which defines the slope of the sliding surface.

Definition A sliding motion will appear on the sliding manifold $S_c(\tau_n, \tau) = \tau_c(t) = 0$ after a time t_h, if the condition $S_c(t)\dot{S}_c(t) = \tau_c(t)\dot{\tau}_c(t) < 0$ is satisfied for all t in some nontrivial semi-open subinterval of time of the form $[t, t_h) \subset (0, t_h)$.

3.1.2 Proposed Parameter Update Rules for the T2FNN

Theorem 1 *If the adaptation laws for the parameters of the considered T2FNN are chosen as [28]:*

$$\dot{\underline{c}}_{1i} = \dot{\overline{c}}_{1i} = \dot{c}_{1i} = \dot{x}_1 \tag{12}$$

$$\dot{\underline{c}}_{2j} = \dot{\overline{c}}_{2j} = \dot{c}_{2j} = \dot{x}_2 \tag{13}$$

$$\dot{\underline{d}}_{1i} = \underline{\mu}_{1i}\frac{-\alpha \underline{d}_{1i}^2}{x_1 - c_{1i}}sgn(\tau_c)sgn\left(\frac{x_1 - c_{1i}}{\underline{d}_{1i}}\right) \tag{14}$$

$$\dot{\overline{d}}_{1i} = \overline{\mu}_{1i}\frac{-\alpha \overline{d}_{1i}^2}{x_1 - c_{1i}}sgn(\tau_c)sgn\left(\frac{x_1 - c_{1i}}{\overline{d}_{1i}}\right) \tag{15}$$

$$\dot{\underline{d_{2j}}} = \underline{\mu_{2j}} \frac{-\alpha \underline{d_{2j}}^2}{\underline{x_2 - c_{2j}}} sgn(\tau_c)sgn\left(\frac{x_2 - \underline{c_{2j}}}{\underline{d_{2j}}}\right) \tag{16}$$

$$\overline{\dot{d}_{2j}} = \overline{\mu_{2j}} \frac{-\alpha \overline{d_{2j}}^2}{\overline{x_2 - c_{2j}}} sgn(\tau_c)sgn\left(\frac{x_2 - \overline{c_{2j}}}{\overline{d_{2j}}}\right) \tag{17}$$

$$\dot{f}_{ij} = -\frac{\left(q(t)\widetilde{W}_{ij} + \left(1 - q(t)\right)\widetilde{\overline{W}}_{ij}\right)\alpha sgn(\tau_c)}{\left(q(t)\widetilde{\underline{W}} + \left(1 - q(t)\right)\widetilde{\overline{W}}\right)^T \left(q(t)\widetilde{\underline{W}} + \left(1 - q(t)\right)\widetilde{\overline{W}}\right)} \tag{18}$$

$$\dot{q}(t) = -\frac{\alpha sgn(\tau_c)}{F(\widetilde{\underline{W}} - \widetilde{\overline{W}})^T} \tag{19}$$

where α is a sufficiently large positive design constant satisfying the inequality below:

$$\alpha > B_{\dot{\tau}} \tag{20}$$

Then, given an arbitrary initial condition $\tau_c(0)$, the learning error $\tau_c(t)$ will converge to zero within a finite time t_h.

Proof The reader is referred to [28].

The relation between the sliding line S_p and the zero adaptive learning error level S_c is determined by the following equation:

$$S_c = \tau_c = k_D\dot{e} + k_Pe = k_D\left(\dot{e} + \frac{k_p}{k_D}e\right) = k_DS_p \tag{21}$$

The tracking performance of the feedback control system can be analyzed by introducing the following Lyapunov function candidate:

$$V_p = \frac{1}{2}S_p^2 \tag{22}$$

Theorem 2 *If the adaptation strategy for the adjustable parameters of the T2FNN is chosen as in (12)–(19), then the negative definiteness of the time derivative of the Lyapunov function in (22) is ensured.*

Proof The reader is referred to [28].

Remark The obtained result means that, assuming that the SMC task is achievable, using τ_c as a learning error for the T2FNN together with the adaptation laws (12)–(19) enforces the desired reaching mode followed by a sliding regime for the system under control.

3.1.3   Experimental Results for the Model-Free Controller

Even if such a trajectory is not common in a typical agricultural operation, an 8-shaped time-based trajectory is preferred in this investigation. The simple reason behind such a trajectory is to be able to elaborate the performance of the intelligent model-free controller both for straight and curved lines.

As can be seen from Figs. 5a–c, the proposed model-free control scheme consisting of a T2FNN working in parallel with a conventional controller gives a better trajectory following accuracy than the one where only a PD controller acts alone. One can claim that the same performance can be obtained by further tuning the conventional controller. However, when there exists more than one subsystem as well as the changing parameters of the system model and the variations in working conditions (soil and crop variability), this task is not straightforward.

In order to show the adaptability capability of the proposed scheme, we show the difference between the first, second and thirds turns for different controllers in Fig. 5a. As it is expected, when the PD controller acts alone, its performance does not change from the first turn to the consequent turns. Thanks to the learning capability, the T2FNN working in parallel with a PD controller gives a better performance in its second and third turns. The results in Fig. 5b show performance improvement of approximately 30% in the case of having a PD controller working in parallel with the T2FNN.

The controller signals coming from the PD controller and the T2FNN can be seen in Fig. 5c. In its first turn, the dominating control signal is coming from the PD controller. In its second turn (starting from 120th s), the T2FNN is able to take over the control, thus becoming the leading controller. Every time there is a change in the reference signal, after a finite time, the output of the PD controller again tends to go to zero. As can be seen from Fig. 5d, the T2FNN significantly increases the control accuracy of the yaw dynamics of the system.

3.1.4   Discussions for Model-Free Control

The real-time test results are promising in a way that when the system is controlled by using a conventional controller in parallel with a T2FNN, the accuracy of the overall controller increases. In this method, the conventional controller is responsible for the stability of the system in the beginning of the learning process. After the learning process starts, the T2FNN controller learns the system dynamics and takes the responsibility of controlling the system gradually. In other words, there is no need to well-tune the conventional controller. It is to be noted that in complex mechatronic systems where there exist more than one subsystem, well-tuning of different controllers on different subsystems is a tedious work.

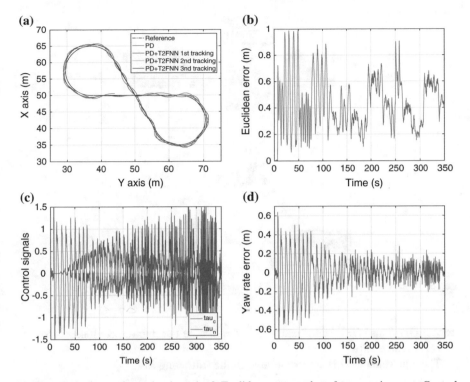

Fig. 5 **a** Reference and actual trajectories **b** Euclidean error to the reference trajectory **c** Control inputs **d** Yaw rate error

3.2 Model-Based Learning: Model Predictive Control-Moving Horizon Estimation Framework

Nonlinear model predictive control and nonlinear moving horizon estimation framework is illustrated in Fig. 6 and a system model is required to design this framework. The equations for the tractor are written as follows:

$$\dot{x} = \mu v \cos(\psi)$$
$$\dot{y} = \mu v \sin(\psi)$$
$$\dot{\psi} = \frac{\mu v \tan(\kappa \delta)}{L}$$
$$\dot{v} = -\frac{v}{\tau} + \frac{K}{\tau} HP \tag{23}$$

where x, y and ψ denote respectively the positions and yaw angle of the tractor while v denotes the speed. The steering angle and hydrostat position are respectively

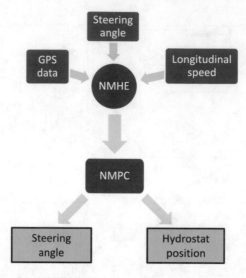

Fig. 6 Block diagram of the NMHE-NMPC framework

denoted by δ and HP. Additionally, μ and η denote the traction coefficients for the wheel and side slips.

The equations in (23) are formulated in the following form;

$$\dot{\xi} = f(\xi, u, p) \tag{24}$$

$$y = h(\xi, u, p) \tag{25}$$

with

$$\xi = \begin{bmatrix} x \ y \ \psi \ v \end{bmatrix}^T \tag{26}$$

$$u = \begin{bmatrix} \delta \ HP \end{bmatrix}^T \tag{27}$$

$$p = \begin{bmatrix} \mu \ \eta \end{bmatrix}^T \tag{28}$$

$$y = \begin{bmatrix} x \ y \ v \ \delta \ HP \end{bmatrix}^T \tag{29}$$

where ξ, u, p and y denote respectively the vectors of state, input, parameter and output of the system. The measured physical parameter is: $L = (1.4\,\text{m})$, and the identified parameters are [38]: $\tau = 2.05$ and $K = 0.016$ for the speed model while the engine speed is at 2500 RPM.

3.2.1 Nonlinear Moving Horizon Estimation

In advanced model-based control structures, learning phenomena are required and realized through online parameter estimation as they make use of the system model

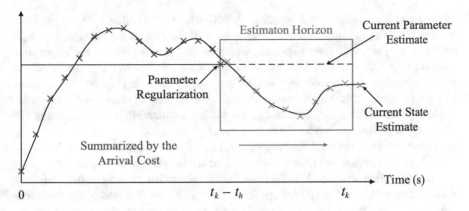

Fig. 7 Illustration for the concept of NMHE

to generate control signals, and have to deal with uncertain and varying process conditions. Therefore, it is inevitable to use adaptive models. In this study, nonlinear moving horizon estimation method has been chosen as a state and parameter estimation algorithm because it considers the state and parameter estimation within the same problem and allows to incorporate constraints both on states and parameters. NMHE is illustrated in Fig. 7 and formulated in (30).

$$
\min_{\xi(.),dp,u(.)} \left\| \begin{matrix} \hat{\xi} - \xi(t_k - t_h) \\ \hat{p} - p \end{matrix} \right\|^2_{V_s} + \int_{t_k - t_h}^{t_k} \left(\|y_m - y(t)\|^2_{V_y} \right) dt
$$

$$
\begin{aligned}
\text{subject to} \quad & \dot{\xi}(t) = f\big(\xi(t), u(t), p\big) \\
& y(t) = h\big(\xi(t), u(t), p\big) \\
& \xi_{min} < \xi < \xi_{max} \\
& p_{min} < p < p_{max} \qquad \forall t \in [t_k, t_{k+1}]
\end{aligned}
\tag{30}
$$

In practice, only state estimation is not enough to know the system behaviour when uncertain systems are considered. Hence, parameter estimation is required to determine unmeasurable parameters. A parametric least square estimation subject to the system model and/or boundary conditions has been studied. There are many software packages to solve optimization problems for offline parameter estimation [29].

Two approximations have been proposed for online parameter estimation which is necessary simultaneously with state estimation to find system behavior accurately. In the first choice, model parameters are assumed as so-called "random constants" represented by a differential equation $\dot{\xi}_p = 0$ with initial value $\xi_p(t_k) = p_k$. This approach results in time-invariant parameters over the estimation horizon. If jumps or drifts for parameters are expected, which is the case in practice under varying working conditions, a model bias would occur. As a solution to the jump and drifts problem, the model parameters must be assumed as time-varying. Model parameters are assumed as so-called "random walk" by a differential equation $\dot{\xi}_p = \frac{dp}{\Delta t}$ with sampling time Δt and initial value $\xi_p(t_k) = p_k$. It is assumed that the parameters are time-varying Gaussian random variables in the arrival cost.

The reference estimated values $\hat{\xi}(t_k - t_h)$ and $\hat{p}$ are taken from the solution of NMHE at the previous estimation instant. The arrival cost matrix V_s has been chosen as a so-called smoothed EKF-update based on sensitivity information obtained while solving the previous NMHE problem [30]. The contributions of the past measurements to the inverted Kalman covariance V_s are downweighted by a process noise covariance matrix D_{update} in (32) [31–35].

The adaptive kinematic model presented in (23) is used in the NMHE design for the state and parameter estimation. The NMHE problem is solved at each sampling time with the following constraints on the parameters:

$$0.25 \leq \mu \leq 1$$
$$0.25 \leq \eta \leq 1 \tag{31}$$

The standard deviations of the measurements have been set to $\sigma_x = \sigma_y = 0.03$ m, $\sigma_v = 0.1$ m/s, and $\sigma_\delta = 0.0175$ rad based on the information obtained from the real-time experiments. Thus, the following weighting matrices V_y, and D_{update} have been used in the NMHE implementation:

$$\begin{aligned} V_y &= diag(\sigma_x, \sigma_y, \sigma_v, \sigma_\delta)^{-1} \\ &= diag(0.03, 0.03, 0.1, 0.0175)^{-1} \end{aligned} \tag{32}$$
$$\begin{aligned} D_{update} &= diag(x, y, \psi, \mu, \eta, v) \\ &= diag(10.0, 10.0, 0.1, 0.25, 0.25, 0.1)^{-1} \end{aligned} \tag{33}$$

The estimation horizon t_h is set to 3 s.

3.2.2   Nonlinear Model Predictive Control

In this study, an NMPC formulation at each sampling time t is considered in the following form:

$$\min_{\xi(.),u(.)} \int_{t_k}^{t_k+t_h} \left(\|\xi_r(t) - \xi(t)\|_Q^2 + \|u_r(t) - u(t)\|_R^2 \right) dt$$
$$+ \|\xi_r(t_k + t_h) - \xi(t_k + t_h)\|_S^2$$

$$\text{s.t.} \quad \xi(t_k) = \hat{\xi}(t_k)$$
$$\dot{\xi}(t) = f\left(\xi(t), u(t), p\right)$$
$$\xi_{min} \le \xi(t) \le \xi_{max}$$
$$u_{min} \le u(t) \le u_{max} \quad \forall t \in [t_k, t_k + t_h]$$

$$(34)$$

where the first and last parts are called the stage cost and the terminal penalty enforced the stability of NMPC in [36] in which $Q \in \mathbb{R}^{n_\xi \times n_\xi}$, $R \in \mathbb{R}^{n_u \times n_u}$ and $S \in \mathbb{R}^{n_\xi \times n_\xi}$ are symmetric positive definite weighting matrices, ξ_r and u_r denote respectively the references for the states and inputs, ξ and u denote respectively the states and inputs, t_k denotes the current time, t_h denotes the prediction horizon. $\hat{\xi}(t_k)$ denotes the estimated state vector by the NMHE, ξ_{min}, ξ_{max}, u_{min} and u_{max} denote respectively the upper and lower constraints on the state and input. The terminal constraints are denoted by $\xi(t_k + t_h)_{min}$ and $\xi(t_k + t_h)_{max}$. The first sample of $u(t)$, $u(t, \xi(t)) = u^*(t_k)$, is applied to the system and the NMPC problem is solved again over a moved horizon for the subsequent sampling time.

The constraints on the inputs, the steering angle and hydrostat position references, are written:

$$-35 \deg \le \delta(t) \le 35 \deg$$
$$0\% \quad \le HP(t) \le 100\%$$

$$(35)$$

The references for the state and inputs are written:

$$\xi_r = (x_r, y_r, \psi_r, v_r)^T$$
$$u_r = (\delta_r, HP_r)^T$$

$$(36)$$

The inputs references are the last measured steering angle and hydrostat position while the states references are relied on the reference trajectory to be tracked.

The weighting matrices Q, R and S have been written:

$$Q = diag(1, 1, 0, 0)$$
$$R = diag(5, 5)$$
$$S = diag(10, 10, 0, 0)$$

$$(37)$$

The prediction horizon t_h is set to 3 s.

3.2.3 Experimental Results for the Model-Based Controller

Throughout the experiments, the tractor has faced with uneven terrain, and been succeed in staying on-track for the NMHE-NMPC framework as shown in Fig. 8a. The sampling time of the frameworks is 200 ms in real-time. The Euclidean error for the tractor is shown in Fig. 8b. The mean values of the Euclidean error of the tractor have been obtained 18.16 cm for the straight lines while 52.02 cm for the curved lines. It is observed that the trajectory tracking error for the system for straight lines has been less than the one for the curved lines as shown in Fig. 8b.

The outputs of the controller, which are the steering angle reference for the tractor (δ^t), and the hydrostat position (HP) reference as illustrated in Fig. 8c. As seen in this figure, the control signals stay within the bounds. Moreover, the estimated traction parameters by the NMHE are shown in Fig. 8d. The estimates stay within the bounds.

For auto generation of the C codes, an open source software is preferred: the ACADO [37] code generation tool. This too can be used to solve the constrained nonlinear optimization problems in the NMPC and NMHE. First, we have created the

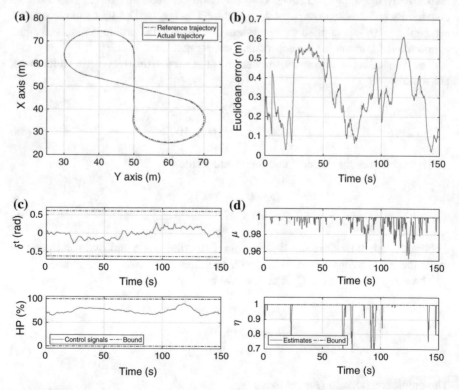

Fig. 8 **a** Reference and actual trajectories **b** Euclidean error to the reference trajectory **c** Control inputs **d** Traction parameters

C codes by using the ACADO, which is then converted into a .dll file to be used in LabVIEW. Detailed information on the ACADO code generation tool can be found in [29, 37].

3.3   Overall Comparison of the Model-Based and Model-Free Learning Control

Table 1 is a candidate guideline to choose an appropriate control algorithm for the control of agricultural robotic systems which are generally complex mechatronic systems. According to observations, when the model of the system as well as the interactions between the subsystems are precisely modeled, a model-based controller (MPC-MHE framework) is preferable. This advanced control framework is not only very accurate but also robust. Moreover, there are some open course fast solvers, such as Acado toolkit, which generates C/C++ codes for a real-time implementation. What is more, although early MPC applications were restricted only to slow systems that long computation times could be tolerated, recent progress in microprocessor technology has motivated applications of MPC for fast dynamic systems, such autonomous vehicles.

However, if the modeling of the system is challenging or unfeasible, model-free control algorithms can be used even if it might be difficult to prove their stability. In addition to their instability problems, pure model-free methods may be unstable in the beginning of the experiment depending on the initial weights. If the system dynamics are fast, this may cause serious problems, such as fast vehicles or unmanned aerial vehicles. In order to make sure that the system is stable in the beginning of the learning process, an alternative method, which is the combination of a conventional controller and an intelligent structure. This fusion is called feedback error learning, which is also promising in real time if there is no precise model at hand. These controllers have the ability of learning throughout the operation if an appropriate optimization algorithm is used.

No matter the model-free controller is a pure model-free or a feedback error learning-based controller, another prominent feature of them is that the time spent for modeling does not exist for model-free controllers. It is to be noted that the modeling stage may take more time than designing of a controller in the case of having a model-based controller. In particular, in addition to its nonlinearities, if the system has dead-zones and hysteresis, modeling of the system is a very tedious work [38]. These challenging systems include, but are not limited to, electro-hydraulic actuators and valves, diesel engines and pneumatic actuators.

Table 1 Guidelines for the selection of the best modeling and control approach for the complex mechatronic systems

Model-based techniques	
When?	Interactions are known accurately
Why?	Allows to design the controller analytically and to prove the stability of the overall system
Why not?	In practical applications, the interactions are not so easy to be modelled
Pure model-free techniques	
When?	Interactions are difficult to be known
Why?	No need the mathematical model of the system to be controlled
Why not?	Impossible to prove the stability of the overall system and impossible to calculate the parameters of the controller analytically
Feedback error learning	
When?	A conventional controller to guarantee the stability of the plant
Why?	After the intelligent controller has learned the system dynamics, it takes the responsibility of controlling the system
Why not?	The stability of the overall system may be challenging to be shown

4 Conclusions

A fully autonomous tractor-trailer system is designed and prototyped by using off the shelf components. The system is able to follow both straight line and curved line trajectories with a satisfactory accuracy. Both model-based and model-free controllers are designed to navigate the system, and their performances are compared and contrasted. According to the real-time results, when the model of the system as well as the interactions between the subsystems are precisely modeled, a model-based controller is preferable. On the other hand, a model-free controller is preferable if the mathematical model of the system is challenging or unfeasible. As a model-free control algorithm, type-2 fuzzy logic controllers are able to learn the systems dynamics online, and have the ability to control the system with a limited information about the system. As a parameter update algorithm, a sliding mode control theory-based learning algorithm is preferred which need neither partial derivatives nor matrix inversions. These features make the learning algorithm not only robust but also computationally efficient which is a big advantage in real-time implementations where the computation power is limited.

References

1. S. Riehl, M. Zeidi, N.J. Conard, Emergence of agriculture in the foothills of the Zagros mountains of Iran. Science **341**(6141), 65–67 (2013). https://doi.org/10.1126/science.1236743, http://www.sciencemag.org/content/341/6141/65.abstract

2. S. Riehl, M. Benz, N. Conard, H. Darabi, K. Deckers, H. Nashli, M. Zeidi-Kulehparcheh, Plant use in three Pre-Pottery Neolithic sites of the northern and eastern Fertile Crescent: a preliminary report. Veg. Hist. Archaeobotany **21**(2), 95–106 (2012). https://doi.org/10.1007/s00334-011-0318-y
3. G. Willcox, The roots of cultivation in Southwestern Asia. Science **341**(6141), 39–40 (2013). https://doi.org/10.1126/science.1240496, http://www.sciencemag.org/content/341/6141/39.short
4. J. Oeppen, J.W. Vaupel, Broken limits to life expectancy. Science **296**(5570), 1029–1031 (2002). https://doi.org/10.1126/science.1069675, http://www.sciencemag.org/content/296/5570/1029.short
5. F. Sistler, Robotics and intelligent machines in agriculture. IEEE J. Robot. Autom. **3**(1), 3–6 (1987). https://doi.org/10.1109/JRA.1987.1087074
6. R. Harrell, Economic analysis of robotic citrus harvesting in Florida. Trans. ASAE **30**(2), 298–304 (1987)
7. N. Irie, N. Taguchi, T. Horie, T. Ishimatsu, Asparagus harvesting robot coordinated with 3-D vision sensor, in *Industrial Technology, ICIT 2009. IEEE International Conference on*, pp. 1–6 (2009). https://doi.org/10.1109/ICIT.2009.4939556
8. H. Shen, D. Zhao, W. Ji, Y. Chen, J. Lv, Research on the Strategy of Advancing Harvest Efficiency of Fruit Harvest Robot in the Oscillation Conditions, in *Intelligent Human-Machine Systems and Cybernetics (IHMSC), 2011 International Conference on*, vol. 1 (2011), pp. 215–218. https://doi.org/10.1109/IHMSC.2011.58
9. T.T. Nguyen, K. Vandevoorde, N. Wouters, E. Kayacan, J.G.D. Baerdemaeker, W. Saeys, Detection of red and bicoloured apples on tree with an RGB-D camera. Biosyst. Eng. (2016). https://doi.org/10.1016/j.biosystemseng.2016.01.007, http://www.sciencedirect.com/science/article/pii/S1537511016000088
10. A. Gollakota, M.B. Srinivas, Agribot A multipurpose agricultural robot, in *2011 Annual IEEE India Conference* (2011), pp. 1–4. https://doi.org/10.1109/INDCON.2011.6139624
11. D.A. Johnson, D.J. Naffin, J.S. Puhalla, J. Sanchez, C.K. Wellington, Development and implementation of a team of robotic tractors for autonomous peat moss harvesting. J. Field Robot. **26**(6–7), 549–571 (2009), https://doi.org/10.1002/rob.20297
12. D. Bochtis, C. Sorensen, S. Vougioukas, Path planning for in-field navigation-aiding of service units. Comput. Electron. Agric. **74**(1), 80–90 (2010). https://doi.org/10.1016/j.compag.2010.06.008, http://www.sciencedirect.com/science/article/pii/S0168169910001250
13. N. Noguchi, J. Will, J. Reid, Q. Zhang, Development of a master–slave robot system for farm operations. Comput. Electron. Agric. **44**(1), 1–19 (2004). https://doi.org/10.1016/j.compag.2004.01.006, http://www.sciencedirect.com/science/article/pii/S0168169904000316
14. C. Cariou, R. Lenain, B. Thuilot, M. Berducat, Automatic guidance of a four-wheel-steering mobile robot for accurate field operations. J. Field Robot. **26**(6–7), 504–518 (2009). https://doi.org/10.1002/rob.20282
15. J. Backman, T. Oksanen, A. Visala, Navigation system for agricultural machines: Nonlinear Model Predictive path tracking. Comput. Electron. Agric. **82**, 32–43 (2012). https://doi.org/10.1016/j.compag.2011.12.009, http://www.sciencedirect.com/science/article/pii/S0168169911003218
16. H. Hagras, M. Colley, V. Callaghan, M. Carr-West, Online learning and adaptation of autonomous mobile robots for sustainable agriculture. Auton. Robots **13**(1), 37–52 (2002). https://doi.org/10.1023/A:1015626121039
17. M. Bergerman, S.M. Maeta, J. Zhang, G.M. Freitas, B. Hamner, S. Singh, G. Kantor, Robot farmers: autonomous orchard vehicles help tree fruit production. IEEE Robot. Autom. Mag. **22**(1), 54–63 (2015). https://doi.org/10.1109/MRA.2014.2369292
18. F. Rovira-Ms, I. Chatterjee, V. Siz-Rubio, The role of GNSS in the navigation strategies of cost-effective agricultural robots. Comput. Electron. Agric **112**, 172–183 (2015). https://doi.org/10.1016/j.compag.2014.12.017, http://www.sciencedirect.com/science/article/pii/S0168169914003275. Precision Agriculture

19. A.V. Topalov, E. Kayacan, Y. Oniz, O. Kaynak, Adaptive neuro-fuzzy control with sliding mode learning algorithm: Application to Antilock Braking System, in *2009 7th Asian Control Conference* (2009), pp. 784–789

20. H. Li, C. Wu, P. Shi, Y. Gao, Control of nonlinear networked systems with packet dropouts: interval type-2 fuzzy model-based approach. IEEE Trans. Cybern. **45**(11), 2378–2389 (2015). https://doi.org/10.1109/TCYB.2014.2371814

21. H.K. Lam, H. Li, C. Deters, E.L. Secco, H.A. Wurdemann, K. Althoefer, Control design for interval type-2 fuzzy systems under imperfect premise matching. IEEE Trans. Ind. Electron. **61**(2), 956–968 (2014). https://doi.org/10.1109/TIE.2013.2253064

22. O. Castillo, P. Melin, A review on interval type-2 fuzzy logic applications in intelligent control. Inf. Sci. **279**, 615–631 (2014). https://doi.org/10.1016/j.ins.2014.04.015, http://www.sciencedirect.com/science/article/pii/S0020025514004629

23. J. Mendel, H. Hagras, W.W. Tan, W.W. Melek, H. Ying, *Introduction To Type-2 Fuzzy Logic Control: Theory and Applications*, 1st edn. (Wiley-IEEE Press, 2014)

24. E. Kayacan, O. Kaynak, R. Abiyev, J. Trresen, M. Hvin, K. Glette, Design of an adaptive interval type-2 fuzzy logic controller for the position control of a servo system with an intelligent sensor, in *International Conference on Fuzzy Systems* (2010), pp. 1–8. https://doi.org/10.1109/FUZZY.2010.5584629

25. E. Kayacan, E. Kayacan, M.A. Khanesar, Identification of nonlinear dynamic systems using type-2 fuzzy neural networks—A novel learning algorithm and a comparative study. IEEE Trans. Industr. Electron. **62**(3), 1716–1724 (2015). https://doi.org/10.1109/TIE.2014.2345353

26. E. Kayacan, W. Saeys, E. Kayacan, H. Ramon, O. Kaynak, Intelligent control of a tractor-implement system using type-2 fuzzy neural networks. *2012 IEEE International Conference on Fuzzy Systems* (2012), pp. 1–8. https://doi.org/10.1109/FUZZ-IEEE.2012.6250790

27. A. Dutta, Y. Zhong, B. Depraetere, K.V. Vaerenbergh, C. Ionescu, B. Wyns, G. Pinte, A. Nowe, J. Swevers, R.D. Keyser, Model-based and model-free learning strategies for wet clutch control. Mechatronics **24**(8), 1008–1020 (2014). https://doi.org/10.1016/j.mechatronics.2014.03.006, http://www.sciencedirect.com/science/article/pii/S0957415814000622

28. E. Kayacan, E. Kayacan, H. Ramon, O. Kaynak, W. Saeys, Towards agrobots: trajectory control of an autonomous tractor using type-2 fuzzy logic controllers. Mechatron. IEEE/ASME Trans. **20**(1), 287–298 (2015). https://doi.org/10.1109/TMECH.2013.2291874

29. B. Houska, H.J. Ferreau, M. Diehl, ACADO toolkit - An open-source framework for automatic control and dynamic optimization. Optimal Control Appl. Meth. **32**(3), 298–312 (2011)

30. D. Robertson, *Development and Statistical Interpretation of Tools for Nonlinear Estimation* (Auburn University, 1996)

31. T. Kraus, H. Ferreau, E. Kayacan, H. Ramon, J.D. Baerdemaeker, M. Diehl, W. Saeys, Moving horizon estimation and nonlinear model predictive control for autonomous agricultural vehicles. Comput. Electron. Agric. **98**, 25–33 (2013)

32. E. Kayacan, J.M. Peschel, E. Kayacan, Centralized, decentralized and distributed nonlinear model predictive control of a tractor-trailer system: A comparative study. *2016 American Control Conference (ACC)*, (2016), pp. 4403–4408. https://doi.org/10.1109/ACC.2016.7525615

33. E. Kayacan, E. Kayacan, H. Ramon, W. Saeys, Robust tube-based decentralized nonlinear model predictive control of an autonomous tractor-trailer system. IEEE/ASME Trans. Mechatron. **20**(1), 447–456 (2015). https://doi.org/10.1109/TMECH.2014.2334612

34. E. Kayacan, E. Kayacan, H. Ramon, W. Saeys, Learning in centralized nonlinear model predictive control: application to an autonomous tractor-trailer system. IEEE Trans. Control Syst. Technol. **23**(1), 197–205 (2015). https://doi.org/10.1109/TCST.2014.2321514

35. E. Kayacan, E. Kayacan, H. Ramon, W. Saeys, Distributed nonlinear model predictive control of an autonomous tractor–trailer system. Mechatronics **24**(8), 926–933 (2014). https://doi.org/10.1016/j.mechatronics.2014.03.007, http://www.sciencedirect.com/science/article/pii/S0957415814000634

36. D. Mayne, J. Rawlings, C. Rao, P. Scokaert, Constrained model predictive control: stability and optimality. Automatica **36**(6), 789–814 (2000)

37. B. Houska, H.J. Ferreau, M. Diehl, An auto-generated real-time iteration algorithm for nonlinear MPC in the microsecond range. Automatica **47**(10), 2279–2285 (2011)
38. E. Kayacan, E. Kayacan, H. Ramon, W. Saeys, Nonlinear modeling and identification of an autonomous tractor–trailer system. Comput. Electron. Agric. **106**, 1–10 (2014). https://doi.org/10.1016/j.compag.2014.05.002, http://www.sciencedirect.com/science/article/pii/S0168169914001252

Important and Challenging Issues for Interval Type-2 Fuzzy Control Research

Hao Ying

Abstract The author points out three important issues: (1) when should interval type-2 (IT2) fuzzy control be utilized, (2) how to design IT2 fuzzy controllers, and (3) how to analyze IT2 fuzzy controllers. Discussion is focused on application and practicality.

1 Introduction to Interval Type-2 Fuzzy Control

Fuzzy control is the most active and victorious component of fuzzy systems technology. The first fuzzy controller was developed by Professor E. H. Mamdan at University of London in United Kingdom in 1974 [4]. The primary thrust of this novel control paradigm at the time was to utilize human control operator's knowledge and experience to intuitively construct a controller so that the resulting controller is able to emulate human control behavior to a certain extent. Compared to the traditional control paradigm, the advantages of the fuzzy control paradigm are two folds. First, a mathematical model of the system to be controlled is not required, and (2) a satisfactory nonlinear controller can be developed empirically without complicated mathematics. The core value of these advantages is the practicality—real-word systems are nonlinear; accurately modeling them is difficult, costly, and even impossible in most cases. Proper use of fuzzy control can significantly shorten product research and development time with reduced cost. Since mid-1980s, companies around the world have utilized fuzzy control to make better, cheaper, and smarter products. Many of them are commercial products. All these fuzzy controllers are now called type-1 fuzzy controllers when there is a need to differentiate them from type-2 fuzzy controllers. Nevertheless, they are referred in the literature simply as fuzzy controllers because type-2 fuzzy control did not exist yet when the reports were published. Figure 1 illustrates configuration of a typical type-1 fuzzy controller.

H. Ying (✉)
Department of Electrical and Computer Engineering,
Wayne State University, Detroit, MI, USA
e-mail: hao.ying@wayne.edu

© Springer International Publishing AG 2018
R. John et al. (eds.), *Type-2 Fuzzy Logic and Systems*,
Studies in Fuzziness and Soft Computing 362,
https://doi.org/10.1007/978-3-319-72892-6_4

75

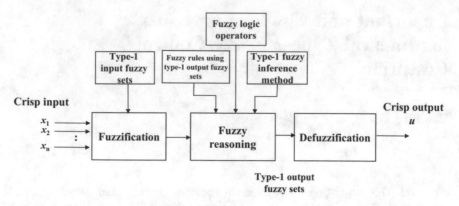

Fig. 1 Configuration of a typical type-1 fuzzy controller

To better reflect complicated nature of expert knowledge, a fuzzy controller may conceivably use a type-2 fuzzy set, which is an extension to a type-1 fuzzy set in that at each value of the universe discourse, the membership value is an interval with another membership function (i.e., secondary membership function) defined over it. A type-2 fuzzy set uses footprint of uncertainty to characterize the region between its upper and lower membership functions. Although the concept of a type-2 fuzzy set was first introduced by Professor L. A. Zadeh in 1975, using it to form a fuzzy inference system is only a relatively recent advance. Professor J. M. Mendel and his coworkers have proposed the first complete type-2 fuzzy inference process, developed various type-2 fuzzy systems, and established their computational principles and foundations since the mid-1990s (e.g., [3, 5, 6]).

With the solid type-2 fuzzy system foundation laid by Mendel and others, researchers extended the notion of fuzzy control to type-2 fuzzy control around the 2000s. The basic idea was to first replace some or all of type-1 fuzzy sets in a fuzzy controllers by (interval) type-2 fuzzy sets, and then added components specific to a type-2 system (e.g., type reducer). Some other modifications were also necessary (e.g., the defuzzification process). Figure 2 shows configuration of a typical type-2 fuzzy controller. The grey boxes spell out the configuration differences between the

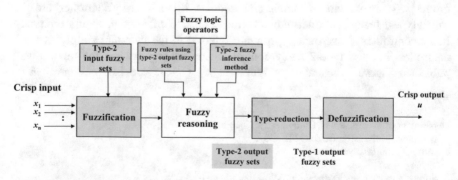

Fig. 2 Configuration of a typical type-2 fuzzy controller

type-2 and type-1 controller configurations in Figs. 1 and 2. Structurally, a type-2 fuzzy controller is more complicated than its type-1 counterpart as the former has more components (e.g., type reducer), more parameters (e.g., footprints of uncertainty of the interval type-2 fuzzy sets), and a more complex inference mechanism.

When the secondary membership function of a type-2 fuzzy set is constant 1, the fuzzy set is an interval type-2 (IT2) fuzzy set. A type-2 fuzzy controller uses IT2 fuzzy sets is called an IT2 fuzzy controller. This chapter focuses on IT2 fuzzy control only as it represents the simplest kind of type-2 fuzzy control and is the most interesting kind at present to the fuzzy control community. Note that an IT2 fuzzy controller degenerates into a type-1 fuzzy controller when footprints of uncertainty of all the type-2 fuzzy sets reduce to 0. Thus, a type-1 fuzzy controller is a special case of this corresponding IT2 fuzzy controller.

2 Research Issue 1: When Should IT2 Fuzzy Control Be Utilized?

Before addressing the issue "when should IT2 fuzzy control be employed to solve a control problem?" let's first discuss the question "when should fuzzy control, type-1 or type-2, be employed to solve a control problem?" Sects. 2.1, 2.2, and 2.4 below are applicable to both type-1 and IT2 fuzzy control.

2.1 Advantages of Fuzzy Control

The biggest advantage of fuzzy control is that it provides an effective and efficient methodology to develop nonlinear controllers without using advanced mathematics. Making a fuzzy controller requires describing human control knowledge/experience linguistically and captures them in the form of fuzzy sets, fuzzy logic operation and fuzzy rules. Fuzzy control can be used to emulate human expert knowledge and experience and is ideal for solving practical problems where imprecision and vagueness are present and verbal description is necessary. Unlike the traditional mathematical-model-based controller design methodology, an explicit system model is not required by fuzzy control. Rather, a system model is implicitly built into fuzzy rules, fuzzy logic operation and fuzzy sets in a vague manner. Fuzzy rules relate input fuzzy sets describing state of output variables of the system to fuzzy controller output. In a sense, fuzzy control combines the system modeling task and the system control task into one task. By avoiding a separate modeling task, which can be more challenging than the control task in many real-world situations, control problems can be solved more efficiently and effectively. Countless applications of fuzzy control around the world have proved this point for type-1 fuzzy control.

Fuzzy control has also created a paradigm for developing nonlinear and multiple-input multiple-output (MIMO) controllers without using sophisticated linear/nonlinear control theory and mathematics. This is in sharp contrast to conventional control technology, especially the nonlinear one. Through manipulating various components of a fuzzy controller, such as the scaling factors, fuzzy sets and fuzzy rules, coupled with computer simulation and/or trial-and-error effort, it is often possible for a non-control professional to build a well-performing fuzzy controller. This advantage makes fuzzy control practical and powerful in solving real-world problems and it explains why (type-1) fuzzy control has especially been popular in industry.

2.2 Disadvantages of Fuzzy Control

A fuzzy controller usually has (far) more design parameters than a comparable conventional controller. To make the matter worse, learning how to construct a good fuzzy controller when the system model is unavailable is, to a large extent, more an art than science. Subsequently, fuzzy controller development may require more tuning and trial-and-error effort. Compared to the industrially dominant PID control that has only three design parameters, the number of design parameters for a fuzzy controller can become overwhelmingly large. They range from the number and shape of input and output fuzzy sets, scaling factors, fuzzy AND and OR operators to fuzzy rules and defuzzifier. Worse yet, there do not exist clear and general relationships between these parameters and controller's performance. The developer need to partially rely on empirical rules of thumb and ad hoc design procedures in the literature to make successful fuzzy control applications. Although there exist a great deal of such knowledge on type-1 fuzzy controllers, it is not sufficient, especially for fuzzy control novices. Fuzzy controllers are nonlinear controllers. As such, the generality of the knowledge is rather limited. Any design and/or tuning procedure can hardly be generalized to cover a broader range of fuzzy control problems. As a result, trial-and-error effort and extensive computer simulation are often necessary. Neither stability nor performance of the fuzzy control system under development can rigorously be guaranteed. This empirical approach, while effective for some applications, is impractical and unsafe for applications in some fields, such as aerospace, nuclear engineering and, particularly, biomedicine.

2.3 Accurate Nonlinear System Models Are Hard and Expansive to Obtain in Practice

Conventional nonlinear control theory is powerful and effective if a nonlinear system model is mathematically available. In order to design a conventional controller for controlling a physical system, the mathematical model of the system is needed. A common form of the system model is differential equation for a continuous-time system or difference equation for a discrete-time system. Strictly speaking, all physical systems are nonlinear. Unless physical insight and the laws of physics can be applied, establishing an accurate nonlinear model using measurement data and system identification methods is difficult in practice.

For any dynamic system modeling problems, linear or nonlinear, two tasks need to be accomplished. The first task is model structure identification, and the second model parameter identification. These tasks are relatively easier for linear system modeling as there have already existed a set of popular linear model structures to choose from, which include AR (Auto Regressive), ARX (Auto Regressive with eXtra input) and ARMA (Auto Regressive Moving Average). They are different types of difference equations and are black-box models. Strictly speaking, a linear system does not exist—a linear model is an approximate model of the nonlinear system valid for a region around one of the system operation points.

Nonlinear system modeling, however, is far more complicated because there exist an infinitive number of possible model structures. Correctly assuming a nonlinear model structure is a hard problem in nonlinear system modeling theory and no general theory exists. Though difficult, different nonlinear system modeling techniques have still been developed, including the Volterra and Wiener theories of nonlinear systems. Such nonlinear system models are black-box models because they only attempt to mimic system's input-output relationship with system measurement data and hence cannot provide any insight on internal structure of the system. Another option is to model a nonlinear system as a (piecewise) linear system. This approach can be over-simplistic in nature and fails to capture diverse and peculiar nonlinear system behaviors, such as limit circles, chaos and bifurcation.

Once the model structure is selected/determined, parameters in the model can be found using system's input-output data and some system optimization procedures (e.g., the least-squares methods), which is the second task.

A linear system model is often adequate for control system development. The whole knowledge base of linear control theory, from linear PID control to modern linear robust control, has been developed based on the notation of linear system models. Once designed, control performance and system stability as well as other properties of the linear control system can usually be examined mathematically. This is because these linear models are difference equations and thus can be analytically analyzed. Whether this linear controller development approach will succeed in practice depends highly on whether the linear model captures the essence of the nonlinear physical system and whether it is a reasonable representation and approximation of the physical system.

In contrast, accurately establishing a nonlinear system model is generally difficult, which significantly limits the application scope of nonlinear control theory.

2.4   When Should Fuzzy Control Be Employed?

There exists literally a countless number of different types of systems in practice. Applicability of fuzzy control, type-1 or type-2, apparently should relate to the strengths and limitations of fuzzy control. In our opinion, fuzzy control is most desirable if (1) mathematical model of the system to be controlled is unavailable but the system is known to be significantly nonlinear, time-varying or with a larger time delay, and/or (2) PID control cannot generate satisfactory system performance.

Given the strengths of fuzzy control, the first criterion is natural and logical. We need to stress the second criterion: It is practically important to know whether PID (including PI or PD control) can solve the control problem of interest before fuzzy control is attempted. PID, PI, PD controllers have been used to control about 90% industrial processes worldwide. PID control techniques are well-developed and numerous control system design and gain tuning methods are available. When the system to be controlled is linear and its mathematical model is available, design and implementation of linear PID control is effective and efficient. Note that using PID control does not necessarily require system model. In the absence of a system model, one can still achieve satisfactory PID control performance in practice by manually tuning, in a trial-and-error fashion, the proportional-gain, integral-gain and derivative-gain. This is true if the system is linear, somewhat nonlinear, or with a mild time delay. Better yet, there exist different types of PID controllers. The most commonly used one is the linear PID controller but often nonlinear ones, such as the anti-windup PID controller, are also employed. Properly adding nonlinearity to linear PID control can lead to desirable control performance. Time has proved that PID control, though simple, is effective and can produce satisfactory results quickly for the majority of control problems, especially those in process control. This is the case even when the system of interest is nonlinear, time-varying or associated with a time delay, as long as they are not too severe.

Fuzzy control should be used, if at least one of the two criteria mentioned above holds. This is the case even if control expert knowledge and experience is unavailable. Practically speaking, it is possible for one to achieve satisfactory fuzzy control of nonlinear systems through extensive computer simulation and trial-and-error effort without expert knowledge. Utilizing available expert knowledge/experience will no doubt reduce development cost and time, particularly when the system is rather complex. But this is not a prerequisite for using fuzzy control.

Even when the system of interest is nonlinear, time-varying or associated with a time delay and its mathematical model is explicitly given, it can often be still advantageous to apply fuzzy control provided that designing an adequate nonlinear controller is more difficult. Unlike linear control theory, there does not exist a general nonlinear control and system theory that is universally applicable to any nonlinear, time-varying or time-delay systems. When a nonlinear system of interest is complicated, or a MIMO one, conventional control theory may be ineffective or even unusable. Furthermore, many of the existing nonlinear control techniques require highly sophisticated control and mathematics background (e.g., differential geometry), which are inaccessible to many of control engineers in the field.

Fuzzy control should not be employed if the system to be controlled is linear, regardless of the availability of its explicit model. For linear systems, there is no advantage to use fuzzy control. PID control and various other types of linear controllers can effectively solve the problem with significantly less effort, time and cost.

In summary, fuzzy control does not and cannot replace conventional control, linear or nonlinear; instead, it complements conventional control rather nicely.

2.5   When Should IT2 Fuzzy Control Be Employed?

In the 1980s, the question "when should fuzzy control be used instead of a conventional controller" was faced by the fuzzy control community. Because the advantages and disadvantages of type-1 fuzzy control relative to those of conventional control were relatively easy to determine and understand, that question was not too difficult to be settled.

A similar question "when should IT2 fuzzy control be used instead of type-1 fuzzy controller" is now waiting the fuzzy control community to answer.

According to Figs. 1 and 2, both type-1 and IT2 fuzzy control methodologies provide a "knowledge engineering" procedure, as opposed to the mathematical approach exclusively adopted in conventional control, to construct $u = f(x_1, x_2, \ldots, x_n)$, where f is a nonlinear and unknown function that represents the control solution being sought. It has been shown that a wide range of type-1 fuzzy controllers are universal approximators in that they can approximate continuous functions arbitrarily well (e.g., [10, 12, 15]), so are various IT2 fuzzy controllers [14]. So, theoretically speaking, IT2 fuzzy control can do whatever type-1 fuzzy control can do, and vice versa.

It should not be difficult to understand that IT2 fuzzy control will not, and cannot, replace either type-1 fuzzy control or conventional control. The three control methodologies are complementary. Arguably, one of the most important research directions is to develop a theory capable of determining whether or not an IT2 fuzzy controller should be used for any given control problem. That is, a theory is needed that can be used ahead of time to determine whether an IT2 fuzzy controller should be employed as opposed to a type-1 fuzzy controller. It is important that such a theory be simple and effective so that it can be used by a control practitioner who may be moderately knowledgeable about type-1 fuzzy control but has little or no knowledge about IT2 fuzzy control (it is not very realistic to assume that someone knowing nothing about type-1 fuzzy control will consider to use IT2 fuzzy control). This theory should not be simulation-based because a system's accurate mathematical model is, realistically speaking, always nonlinear and thus is very difficult to obtain in practice, as we pointed out above. This theory should also not be heavily reliant on trial-and-error effort because such an approach can not only be costly but also risky to use for safety-critical applications (e.g., nuclear industry and clinical medicine).

In practice, IT2 fuzzy control may have to prove its superiority to both type-1 fuzzy control and conventional control for a particular control problem or a particular class of control problems before it will actually be used. Because type-1 fuzzy control and conventional control are able to deliver satisfactory solutions for so many different practical control problems, defining the niche applications that require the distinct merits of IT2 fuzzy control is a critically important but technically challenging area of study. Another important factor that one has to keep in mind is that a real-world control application typically seeks the simplest and least expensive hardware/software solution that satisfies the technical specifications required by the user. This is why PID, PI and PD controllers, with only two or three design parameters, all of which can be tuned manually in an intuitive manner, have become the most popular controllers since their inception, dating back to the pre-electronic period, despite the availability of numerous more advanced and better (at least in theory) controllers developed in the past dozens of years (e.g., optimal controllers and robust controllers).

An IT2 fuzzy controller should not be used unless its added structural complexity and additional design parameters (as compared with a type-1 fuzzy controller) can be reasonably justified by demonstrated significant gains in control performance (e.g., better transition control response and/or more robust performance in the presence of system noise and/or disturbance). Research has been under way to explore when IT2 fuzzy control can bring substantial performance improvement, and more and more results are appearing.

3 Research Issue 2: How to Design IT2 Fuzzy Controllers?

If, for a given practical control problem, it is decided to use an IT2 fuzzy controller instead of a type-1 controller, the next logical issue is how to design it.

Numerous techniques have been developed in literature for analyzing and designing a wide variety of fuzzy control systems of both the Mamdani type and the TSK type. They are mostly for the type-1 fuzzy controllers for now [2], but a growing number of techniques are developed for the IT2 controllers. The literature can be classified into two groups according to methodology: (1) the model-based approach, and (2) the knowledge-based approach, which is a model-free approach. When the model-based approach is used, the precise mathematical model of the system to be controlled must be assumed explicitly available whereas the knowledge-based approach does not make such an assumption. The model of interest should be nonlinear because a practical system is always nonlinear. While the model availability assumption makes theoretical development mathematically tractable and convenient for the model-based approach, it hardly realistically reflect practical constraints. The fact of the matter is this—it is challenging to attain a reasonable nonlinear mathematical model for most systems in the real world. The pitfall of the model availability assumption holds not only for fuzzy control but also equally for conventional control. Emerging in the 1990s, this approach provides mathematical convenience at the cost of practicality. It has produced a large volume of publications; nevertheless, its usefulness in practice has yet to be established. *In short, without knowing the nonlinear model, most, if not all, of the model-based design methods for the type-1 fuzzy controllers are simply inapplicable.*

IT2 fuzzy controllers are nonlinear controllers with complicated input–output relations. They are certainly more complex than their type-1 counterparts in terms of the mathematical input–output relations and the number of design parameters. Consequently, designing an IT2 fuzzy control system is more challenging than designing a type-1 fuzzy control system. As evident by trends in the recent literature, an important research direction is to extend the analysis and design techniques that have been developed for various type-1 fuzzy controllers and systems to IT2 fuzzy controllers and systems. Interestingly, methodologies available for analyzing and designing IT2 fuzzy controllers and systems are fundamentally the same as those utilized for type-1 fuzzy controllers and systems. For example, the Lyapunov approach, which has been widely used for type-1 fuzzy control systems as well as for conventional nonlinear control systems, is the only general tool that has been used for analyzing system stability or designing a stable IT2 fuzzy control system. To date, there exists no other more effective stability approach for IT2 fuzzy control systems. It is presently the most general and best technique available for IT2 fuzzy controllers and systems, and we believe that it will play a crucial role in the development of future IT2 fuzzy control theory. Note, however, that extending the type-1 fuzzy control techniques to cover IT2 fuzzy controllers can be challenging because, generally speaking, an IT2 fuzzy controller is a more complicated nonlinear controller than is a type-1 fuzzy controller.

An IT2 fuzzy controller, like its type-1 counterpart, is presently viewed and treated by most fuzzy control practitioners and theorists as a black-box function generator that is capable of producing a desired nonlinear mapping between input and output of the controller (i.e., $u = f(x_1, x_2, ..., x_n)$ in Fig. 2). The mapping is implicit because $f(x_1, x_2, ..., x_n)$ does not spell out the explicit relationship between the input variables and the output variable. In other words, it shows there is a relationship but does not reveal exactly what it is. When the model-based approach utilizes the implicit $f(x_1, x_2, ..., x_n)$ to develop a controller design method, it treats the fuzzy controller as the black-box function generator. On the other hand, the knowledge-based approach does not start with $f(x_1, x_2, ..., x_n)$. Rather, it relies on a systematic procedure comprising of a number of steps to practically construct $f(x_1, x_2, ..., x_n)$ through manipulating, often in a trial-and-error fashion, fuzzy sets, fuzzy rules, fuzzy inference, and other components. For each component, the developer will face choices. For instance, for input fuzzy sets (i.e., the fuzzy sets for fuzzifying input variables), the developer has to decide how many of them should be used, what type should be used (e.g., triangular vs. Gaussian), and whether a mixture of different types should be used. This is just one of the several components that the developer has to specify (other components include output fuzzy sets, fuzzy rules and defuzzifier). Coupled with computer simulation, this approach often suffices for the practitioner to build a satisfactory fuzzy control system as a solution to the real-world problem at hand. Importantly, this tactic usually works even when the mathematical model of the system is not available. Apart from the approach (model-based or knowledge-based), once built, the fuzzy controller remains a black box in that the explicit expression of $f(x_1, x_2, \ldots, x_n)$ is still unknown. The components work together to generate a value for $f(x_1, x_2, \ldots, x_n)$ for any given value of the input variables. Obviously, the explicit expression of $f(x_1, x_2, \ldots, x_n)$ depends on how the components are selected. The implicit nature of $f(x_1, x_2, \ldots, x_n)$ does not change regardless.

4 Research Issue 3: How to Analyze IT2 Fuzzy Controllers

We call the mapping mentioned above the analytical structure of the fuzzy controller. The model-based approach and the knowledge-based approach of fuzzy control, type-1 or IT2, are in sharp contrast to the conventional control theories. In conventional control, once a controller is chosen by the developer according to the system to be controlled, the controller's analytical structure, linear or nonlinear, is always explicitly ready for analysis and design of the control system. The linear and nonlinear control theories are matured with many time-tested analysis and design schemes. The primary technical difficulty for controller design lies in how to first select or design $f(x_1, x_2, \ldots, x_n)$ and then determine its parameter values based on the given system model so that the designed control system performance will meet the developer's performance specifications. $f(x_1, x_2, \ldots, x_n)$ is explicitly known after the control system design is completed. Control system analysis, stability, control performance, and other system characteristics are analyzed and determined based on both the explicitly $f(x_1, x_2, \ldots, x_n)$ and the system model. To bring fuzzy control to the same level of sophistication and acceptance as the conventional control theories, fuzzy control needs to overcome two hurdles pertinent only to fuzzy control and irrelevant to conventional control. The first hurdle is the unavailability of $f(x_1, x_2, \ldots, x_n)$ in an explicit form after it is designed/constructed, and the second relates to the fundamental question of whether $f(x_1, x_2, \ldots, x_n)$ can be an arbitrary nonlinear function. The second issue, referred to as fuzzy systems as universal approximators in literature, has been extensively addressed for the type-1 fuzzy controllers, but has been investigated for the IT2 controllers only in a rather limited scope [14]. To a large extent, mathematically studying IT2 (or type-1) fuzzy control is inherently even more challenging than studying typical nonlinear control problems. Not explicitly knowing $f(x_1, x_2, \ldots, x_n)$ puts both the model-based and model-free fuzzy control approaches in a disadvantageous position.

Studying the analytical structures of both the controller and the system under control can make it possible for the system analysis and design more precise and effective and less conservative. No matter if an IT2 fuzzy controller is theoretically designed using a model-based scheme or is empirically constructed via a knowledge-based method, revealing controller's analytical structure can be significantly beneficial because one can then:

1. insightfully understand how and why an IT2 fuzzy controller works in the same sense as we understand how a conventional controller functions,
2. find a possible connection between an IT2 fuzzy controller and a conventional controller,
3. explore rigorously the differences between an IT2 fuzzy controller and its type-1 fuzzy controller and their relative merits and pitfalls (e.g., control performance and structural complexity),
4. take advantage of the nonlinear control theory to develop more effective analysis and design methods for IT2 control system as the fuzzy control problem has transformed into a nonlinear control problem, and
5. make IT2 fuzzy control more acceptable to safety-critical fields such as clinical medicine and nuclear industry where people are reluctant to employ a black box as a controller.

We stress that the analytical structure of a fuzzy controller should be investigated in such a way that the structure is sensible in the context of control theory. This is to say that deriving the explicit structure is only a first step, after which the structure should be represented in a form clearly understandable from a control theory standpoint to gain the full potential in system analysis and design.

We derived the first analytical structure of a type-1 fuzzy controller in 1990 [9]. The analytical structures of many other type-1 fuzzy controllers have been reported in the literature since then. The benefits of deriving the analytical structures are well documented in the literature for the type-1 fuzzy controllers. As an example, some type-1 fuzzy controllers have been shown to possess peculiar and interesting structures (e.g., nonlinear PID, PI, or PD controllers with variable gains) [9, 13]. This kind of structural information can be used to guide the parameter-tuning process, thus leading to a significant reduction in trial-and-error effort (e.g., [11, 13]).

Challenges associated with analytical-structure derivation depend on the configuration of the fuzzy controller, in particular, which kind of fuzzy AND operator is used. This is the case for both the type-1 and IT2 fuzzy controllers. The product AND operator and the Zadeh AND operator (i.e., min()) are the only two operators that are employed in fuzzy control. Deriving the analytical structure of a fuzzy controller with the product AND operator is relatively simple; however, a fuzzy controller involving the other operator is far more difficult. Structurally, a IT2 fuzzy controller is more complicated than its type-1 counterpart as the former has more components (e.g., type reducer), more parameters (e.g., footprints of uncertainty of IT2 fuzzy sets), and a more complex inference mechanism.

We revealed first analytical structure of type-2 fuzzy controller which used Zadeh AND operator [1]. Subsequently, the analytical structures of a number of other IT2 fuzzy controllers were exposed. (e.g., [7, 8, 16, 17]). We point out that to study a new class of IT2 fuzzy controllers, an innovative analytical-structure-deriving method must be developed first before their analytical structures can be derived because the existing derivation methods can cover only the controller configurations for which they are developed.

In [1], the analytical structures of two Mamdani IT2 fuzzy PI controllers are derived that use the following identical elements—two interval T2 triangular input fuzzy sets for each of the two input variables, four type-1 singleton output fuzzy sets, Zadeh AND operator, and the center-of-sets type reducer. The difference is that one controller employs the centroid defuzzifier while the other a new defuzzifier called the average defuzzifier (whose advantages are established in the context of the analytical structure study in [1]). The resulting analytical structures are linked to nonlinear control. More specifically, the derivation proves explicitly both controllers to be nonlinear PI (or PD) controllers with variable gains (the expressions are different for the two controllers). The characteristics of the variable gains are analyzed and shown to have the potential to yield improved control performance. Taking advantage of the new knowledge, how to determine and tune the design parameters of the IT2 controllers (there are as many as 11 parameters) even when the mathematical model of the system to be controlled is unknown are discussed.

An innovative technique capable of deriving the analytical structure for a wide class of IT2 Mamdani fuzzy controllers is developed in [16]. The configuration of the controllers is typical and quite general—any number and types of IT2 input fuzzy sets, any number and types of general or IT2 output fuzzy sets, arbitrary fuzzy rules, Zadeh AND operator, the Karnik-Mendel center-of-sets type-reducer, and the centroid defuzzifier. One particularly interesting finding is that the analytical structure of a subset of the IT2 fuzzy controllers is the sum of two nonlinear PI (or PD) controllers, each of which has a variable proportional-gain and a variable integral-gain (or derivative-gain) plus a variable offset if and only if the input fuzzy sets are piecewise linear (e.g., triangular and/or trapezoidal). The sum of the two nonlinear PI (or PD) controllers is a new discovery relative to the literature. As an important benefit of knowing the analytical structure, the IT2 fuzzy controllers can now be treated as variable-gain controllers, rather than black-box controllers. The roles of various parameters, such as the footprints of uncertainty of the IT2 input fuzzy sets, play can be clearly understood from control theory standpoint as opposed to from vague and subjective viewpoint of linguistic knowledge representation. Furthermore, the structure information can be used to facilitate control system design. More concretely, for the fuzzy PI (or PD) controllers, because at the equilibrium point, the variable proportional-gain and integral-gain of the IT2 fuzzy PI (or PD) controller become fixed gains. Therefore, one may apply the linear PI (or PD) controller to the system to be controlled with its mathematical model being assumed to be unknown. Tune the proportional-gain and integral-gain (or derivative-gain) of the linear PI (or PD) controller in a trial-and-error fashion to achieve a reasonable system output performance. The gains of the linear controller can be utilized to calculate the scaling factors of the input and output variables quite easily based on the derived variable gain formulas. The detail on the underlying principle is given in [11].

A long-standing fundamental issue is this: how an IT2 fuzzy set's footprint of uncertainty, a key element differentiating an IT2 controller from a type-1 controller, affects a controller's analytical structure. Absence of a general theory, determining a footprint relies on blind search through the trial-and-error method, which is currently widely adopted in the field. Blind searching of a (high-dimensional) parameter space is not only time consuming but incomprehensive with subpar outcome. We address this issue for a particular class of IT2 TS fuzzy controllers in [17] by first developing an innovative technique for deriving their analytical structures. Analyzing the resulting analytical structures reveals the role of the footprints of uncertainty in shaping the structures. Specifically, it is mathematically proven that under certain conditions, the larger the footprints, the more the IT2 controllers resemble linear or piecewise linear controllers. When the footprints are at their maximum, the IT2 controllers actually become linear or piecewise linear controllers. That is to say the smaller the footprints, the more nonlinear the controllers. The most nonlinear IT2 controllers are attained at zero footprints, at which point the IT2 controllers become type-1 controllers. This finding implies that sometimes if strong nonlinearity is most important and desired, one should consider using a smaller footprint or even just a type-1 fuzzy controller. This study exemplifies the importance of investigating analytical structure of an IT2 fuzzy controller because availability of such structure information can lead to comprehensive and insightful analysis and understanding of an IT2 fuzzy controller.

References

1. X.Y. Du, H. Ying, Derivation and analysis of the analytical structures of the interval type-2 fuzzy PI and PD controllers. IEEE Trans. Fuzzy Syst. **8**, 802–814 (2010)
2. G. Feng, A survey on analysis and design of model-based fuzzy control systems. IEEE Trans. Fuzzy Syst. **14**, 676–697 (2006)
3. N.N. Kamik, J.M. Mendel, Q. Liang, Type-2 fuzzy logic systems. IEEE Trans. Fuzzy Syst. **7** (6), 643–658 (1999)
4. E.H. Mamdani, Application of fuzzy algorithms for simple dynamic plant. Proc. IEE **121**, 1585–1588 (1974)
5. J.M. Mendel, *Uncertain Rule-Based Fuzzy Logic Systems: Introduction and new directions* (Prentice Hall, USA, 2000)
6. J.M. Mendel, Advances in type-2 fuzzy sets and systems. Inf. Sci. **177**, 84–110 (2007)
7. J.M. Mendel, H. Hagras, W. W. Tan, W. Melek, and H. Ying, *Introduction to Type-2 Fuzzy Logic Control: Theory and Applications* (IEEE Press and Wiley, Inc., 2014)
8. M. Nie, W.W. Tan, Analytical structure and characteristics of symmetric Karnik-Mendel type-reduced interval type-2 fuzzy PI and PD controllers. IEEE Trans. Fuzzy Syst. **20**(3), 416–430 (2012)
9. H. Ying, W. Siler, J.J. Buckley, Fuzzy control theory: a nonlinear case. Automatica **26**, 513–520 (1990)
10. H. Ying, Sufficient conditions on general fuzzy systems as function approximators. Automatica **30**, 521–525 (1994)
11. H. Ying, Practical design of nonlinear fuzzy controllers with stability analysis for regulating processes with unknown mathematical models. Automatica **30**, 1185–1195 (1994)

12. H. Ying, Sufficient conditions on uniform approximation of multivariate functions by general Takagi-Sugeno fuzzy systems with linear rule consequent. IEEE Trans. Syst. Man Cybern. **28**, 515–520 (1998)
13. H. Ying, *Fuzzy Control and Modeling: Analytical Foundations and Applications* (IEEE Press, 2000)
14. H. Ying, A sufficient condition on a general class of interval type-2 Takagi-Sugeno fuzzy systems with linear rule consequent as universal approximators. J. Intel. Fuzzy Syst. **29**, 1219–1227 (2015)
15. X.-J. Zeng, M.G. Singh, Approximation properties of fuzzy systems generated by the min inference. IEEE Trans. Syst. Man Cybern. **26**, 187–193 (1996)
16. H.B. Zhou, H. Ying, A method for deriving the analytical structure of a broad class of typical interval type-2 Mamdani fuzzy controllers. IEEE Trans. Fuzzy Syst. **21**, 447–458 (2013)
17. H.B. Zhou, H. Ying, Deriving and analyzing analytical structures of a class of typical interval type-2 TS fuzzy controllers. IEEE Trans. Cybern. **99**, 1–12 (2016)

Type-2 Fuzzy Logic in Pattern Recognition Applications

Patricia Melin

Abstract Type-2 fuzzy systems can be of great help in image analysis and pattern recognition applications. In particular, edge detection is a process usually applied to image sets before the training phase in recognition systems. This preprocessing step helps to extract the most important shapes in an image, ignoring the homogeneous regions and remarking the real objective to classify or recognize. Many traditional and fuzzy edge detectors can be used, but it is very difficult to demonstrate which one is better before the recognition results are obtained. In this work we show experimental results where several edge detectors were used to preprocess the same image sets. Each resulting image set was used as training data for a neural network recognition system, and the recognition rates were compared. In this paper we present the advantage of using a general type-2 fuzzy edge detector method in the preprocessing phase of a face recognition system. The Sobel and Prewitt edge detectors combined with GT2 FSs are considered in this work. In our approach, the main idea is to apply a general type-2 fuzzy edge detector on two image databases to reduce the size of the dataset to be processed in a face recognition system. The recognition rate is compared using different edge detectors including the fuzzy edge detectors (type-1, interval, and general type-2 FS) and the traditional Prewitt and Sobel operators.

Keywords Interval type-2 fuzzy systems · Image processing · Pattern recognition · Edge detection

P. Melin (✉)
Tijuana Institute of Technology, Tijuana, Mexico
e-mail: pmelin@tectijuana.mx

© Springer International Publishing AG 2018
R. John et al. (eds.), *Type-2 Fuzzy Logic and Systems*,
Studies in Fuzziness and Soft Computing 362,
https://doi.org/10.1007/978-3-319-72892-6_5

89

1   Introduction

Edge detection is one of the most common approaches to detect discontinuities in gray scale images. Edge detection can be considered an essential method used in the image processing area and can be applied in image segmentation, object recognition systems, feature extraction and target tracking [1].

There are several edge detection methods, which include the traditional ones, such as Sobel [2], Prewitt [3], Canny [1], Robert [4], Kirsch [5], and those based in type-1 [4, 6–8], interval type-2 [9–12] and general fuzzy systems [13, 14]. In Melin et al. [14] and Gonzalez et al. [13], some edge detectors based on GT2 FSs have been proposed. In these works the results achieved by the GT2 FS are compared with others based on a T1 FS and with an IT2 FS. According with the results obtained in these papers, the conclusion is that the edge detector based on GT2 FS is better than an IT2 FS and a T1 FS.

In other works, like in [15], an edge detector based on T1 FS and other IT2 FS are implemented in the preprocessing phase of a face recognition system. According with the recognition rates achieved in this paper the authors conclude that the recognition system has better performance when the IT2 fuzzy edge detector is applied.

In this paper we present a recognition approach illustrated with faces, which is performed with a monolithic neural network. In the methodology, two GT2 fuzzy edge detectors are applied over two face databases. In the first edge detector a GT2 FS is combined with the Prewitt operator and the second with the Sobel operator. The edge datasets achieved by these GT2 fuzzy edge detectors are using as the inputs of the neural network in a face recognition system.

The aim of this work is to show the advantage of using a GT2 fuzzy edge detector in pattern recognition applications. Additionally, make a comparative analysis with the recognition rates obtained by the GT2 against the results achieved in [15] by T1 and T2 fuzzy edge detectors.

The remainder of this paper is organized as follows. Section 2 gives a review of the background on GT2 FS. The basic concepts about Prewitt, Sobel operator, Low-pass filter and high-pass filter are described in Sect. 3. The methodology used to develop the GT2 fuzzy edge detector is explained in Sect. 4. The design of the recognition system based on monolithic neural network is presented in Sect. 5. The recognition rates achieved by the face recognition system and the comparative results are show in Sect. 6. Finally, Sect. 7 offers some conclusions about the results.

2   Overview of General Type-2 Fuzzy Sets

The GT2 FSs have attracted attention from the research community, and have been applied in different areas, like pattern recognition, control systems, image processing, robotics and decision making to name a few [16–21]. It has been demonstrated that a GT2 FS can have the ability to handle great uncertainties.

In the following we present a brief description about GT2 FS theory, which are used in the methodology proposed in this paper.

2.1 Definition of General Type-2 Fuzzy Sets

A General type-2 fuzzy set $(\tilde{A})$ consists of the primary variable x having domain X, the secondary variable u with domain in J_x^u at each $x \in X$. The secondary membership grade $\mu_{\tilde{A}}(x, u)$ is a 3D membership function where $0 \leq \mu_{\tilde{A}}(x, u) \leq 1$ [22–24]. It can be expressed by (1)

$$\tilde{A} = \left\{ ((x, u), \mu_{\tilde{A}}(x, u)) \mid \forall x \in X, \forall u \in J_x^u \subseteq [0, 1] \right\}. \tag{1}$$

The footprint of uncertainty (FOU) of $(\tilde{A})$ is the two-dimensional support of $\mu_{\tilde{A}}(x, u)$ and can be expressed by (2)

$$FOU(\tilde{A}) = \{(x, u) \in X \times [0, 1] \mid \mu_{\tilde{A}}(x, u) > 0\}. \tag{2}$$

2.2 General Type-2 Fuzzy Systems

The general structure of a GT2 FLS is shown in Fig. 1 and this consists of five main blocks, which are the input fuzzification, fuzzy inference engine, fuzzy rule base, type-reducer and defuzzifier [19].

In a GT2 FLS first the fuzzifier process maps a crisp input vector into other GT2 input FSs. In the inference engine the fuzzy rules are combined and provide a

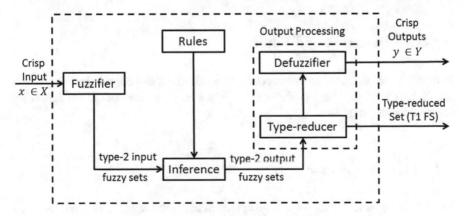

Fig. 1 General type-2 fuzzy logic system

mapping from GT2 FSs input to GT2 FSs output. This GT2 FSs output is reduced to a T1 FSs by the type-reduction process [25, 26].

There are different type-reduction methods, the most commonly used are the Centroid, Height and Center-of-sets type reduction. In this paper we applied Centroid type-reduction. The Centroid definition $C_{\tilde{A}}$ of a GT2 FLS [27–29] is expressed in (3)

$$C_{\tilde{A}} = \{(z_i, \mu(z_i)) | z_i \in \frac{\sum_{i=1}^{N} x_i \theta_i}{\sum_{i=1}^{N} \theta_i},$$
$$\mu(z_i) \in f_{x_1}(\theta_1) \times \ldots \times f_{x_N}(\theta_N), \theta_i \in J_{x_1} \times \ldots \times J_{x_N}\} \tag{3}$$

where θ_i is a combination associated to the secondary degree $f_{x_1}(\theta_1)*\cdots*f_{x_N}(\theta_N)$.

2.3 General Type-2 Fuzzy Systems Approximations

Due to the fact that a GT2 FSs defuzzification process is computationally more complex than T1 and IT2 FSs; several approximation techniques have been proposed, some of them are the zSlices [21, 30] and the α – plane representation [31, 32]. In these two approaches the 3D GT2 membership functions are decompose by using different cuts to achieve a collection of IT2 FSs.

In this paper the defuzzifier process is performed using α – plane approximation, which is defined as follow.

An α-plane for a GT2 FS $\tilde{A}$, is denoted by $\tilde{A}_\alpha$, and it is the union of all primary membership functions of $\tilde{A}$, which secondary membership degrees are higher or equal than α ($0 \leq \alpha \geq 1$) [31, 32]. The α – plane is expressed in (4)

$$\tilde{A}_\alpha = \{(x, u) | \mu_{\tilde{A}}(x, u) \geq \alpha, \forall x \in X, \forall u \in [0, 1]\} \tag{4}$$

3 Edge Detection and Filters

In this Section we introduce some concepts about filters and edge detectors (Prewitt and Sobel) used in image processing areas; since, these are critical for achieving good pattern recognition.

3.1 Prewitt Operator

The Prewitt operator is used for edge detection in digital images. This consists of a pair of 3 × 3 convolution kernels which are defined in (5) and (6) [33].

$$Prewittx = \begin{bmatrix} -1 & -1 & -1 \\ 0 & 0 & 0 \\ 1 & 1 & 1 \end{bmatrix} \tag{5}$$

$$Prewitty = \begin{bmatrix} -1 & 0 & 1 \\ -1 & 0 & 1 \\ -1 & 0 & 1 \end{bmatrix} \tag{6}$$

The kernels in (5) and (6) can be applied separately to the input image (I), to produce separate measurements of the gradient component (7), (8) in horizontally (gx) and vertically orientation (gy) respectively [33].

$$gx = Prewittx * I \tag{7}$$

$$gy = Prewitty * I \tag{8}$$

The gradient components (7) and (8) can be combined together to find the magnitude of the gradient at each point and the orientation of that gradient [2, 34]. The gradient magnitude (G) is given by. (9).

$$G = \sqrt{gx^2 + gy^2} \tag{9}$$

3.2 Sobel Operator

Sobel operator is similar to the Prewitt operator. The only difference is that the Sobel operator use the kernels expressed in (10) and (11) to detect the vertical and horizontal edges.

$$Sobelx = \begin{bmatrix} -1 & -2 & -1 \\ 0 & 0 & 0 \\ 1 & 2 & 1 \end{bmatrix} \tag{10}$$

$$Sobely = \begin{bmatrix} -1 & 0 & 1 \\ -2 & 0 & 2 \\ -1 & 0 & 1 \end{bmatrix} \tag{11}$$

3.3 Low-Pass Filter

Low-pass filters are used for image smoothing and noise reduction; this allows only passing the low frequencies of the image [15]. Also is employed to remove high spatial frequency noise from a digital image. This filter can be implemented by using (12) and the mask (highM) used to obtained the highPF is expressed in (13).

$$lowPF = lowM * I \tag{12}$$

$$lowM = \frac{1}{25} * \begin{bmatrix} 1 & 0 & 0 & 0 & 0 \\ 0 & 1 & 0 & 0 & 0 \\ 0 & 0 & 1 & 0 & 0 \\ 0 & 0 & 0 & 1 & 0 \\ 0 & 0 & 0 & 0 & 1 \end{bmatrix} \tag{13}$$

3.4 High-Pass Filter

High-pass filter only allow the high frequency of the image to pass through the filter and that all of the other frequency are blocked. This filter will highlight regions with intensity variations, such as an edge (will allow to pass the high frequencies) [15]. The high-pass (highPF) filter is implemented by using (14)

$$highPF = highM * I \tag{14}$$

where highM in (14) represents the mask used to obtained the highPF and this is defined by (15)

$$highM = \begin{bmatrix} -1/16 & -1/8 & -1/16 \\ -1/8 & 3/4 & -1/8 \\ -1/16 & -1/8 & -1/16 \end{bmatrix} \tag{15}$$

4 Edge Detection Improved with a General Type-2 Fuzzy System

In our approach two edge detectors are improved, in the first a GT2 FS is combined with Prewitt operator and the second with the Sobel operator. The general structure used to obtain the first GT2 fuzzy edge detector is shown in Fig. 2. The second fuzzy edge detector has a similar structure; we only change the kernel by using the Sobel operators in (10) and (11), which are described in Sect. 3.

The GT2 fuzzy edge detector is calculated as follows. To start, we select a input image (I) of the images database; after that, the horizontal gx (7) and vertical gy (8) image gradients are obtained; moreover, the low-pass (12) and high-pass (14) filters are also applied over (I).

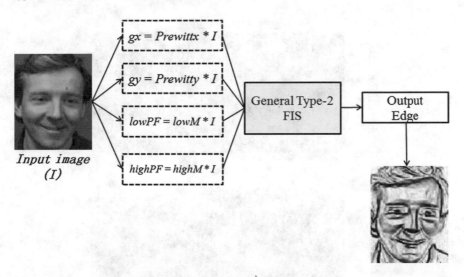

Fig. 2 Edge detector improved with Prewitt operator and GT2 FSs

The GT2 FIS was built using four inputs and one output. The inputs are the values obtained by gx (7), gy (8), $lowPF$ (12) and $highPF$ (14); otherwise, the output inferred represents the fuzzy gradient magnitude which is labeled as Output Edge.

An example of the input and output membership functions used in the GT2 FIS is shown in Figs. 3 and 4 respectively.

In order to objectively compare the performance of the proposed edge detectors against the results achieved in Mendoza [15], we use a similar knowledge base of fuzzy rules; these rules were designed as follows.

1. If (dx is LOW) and (dy is LOW) then (OutputEdge is HIGH)
2. If (dx is MIDDLE) and (dy is MIDDLE) then (OutputEdge is LOW)
3. If (dx is HIGH) and (dy is HIGH) then (OutputEdge is LOW)
4. If (dx is MIDDLE) and (highPF is LOW) then (OutputEdge is LOW)
5. If (dy is MIDDLE) and (highPF is LOW) then (OutputEdge is LOW)
6. If (lowPF is LOW) and (dy is MIDDLE) then (OutputEdge is HIGH)
7. If (lowPF is LOW) and (dx is MIDDLE) then (OutputEdge is HIGH)

5 Face Recognition System Using Monolithic Neural Network and a GT2 Fuzzy Edge Detector

The aim of this work is to apply a GT2 fuzzy edge detector in a preprocessing phase in a face recognition system. In our study case the recognition system is performed using a Monolithic Neural Networks. As already mentioned in Sect. 5, the edge

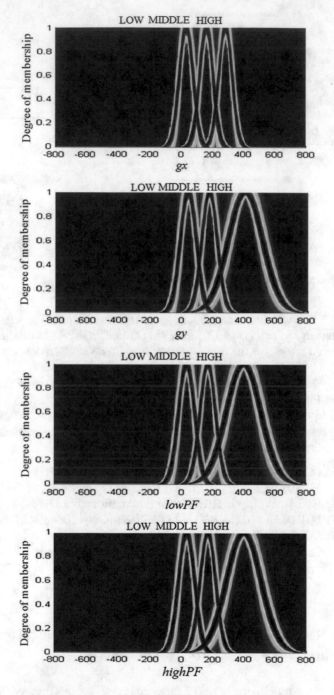

Fig. 3 Input membership functions using in the GT2 FIS

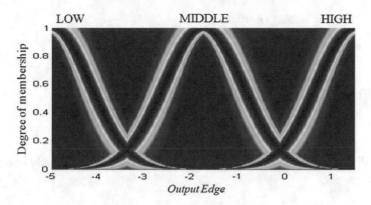

Fig. 4 Output membership function using in the GT2 FIS

detectors were designed using GT2 fuzzy combined with Prewitt and Sobel operator.

In Fig. 4 as an illustration, the general structure used in the proposed face recognition system is shown. The methodology used in the process is summarized in the following steps.

A. **Select the input images database**

In the simulation results two benchmark face databases were selected; in which are included the ORL [35] and the Cropped Yale [36–38].

B. **Applied the edge detection in the input images**

In this preprocessing phase, the two GT2 fuzzy edge detectors described in Sect. 5 were applied on the ORL and Cropped Yale database.

C. **Training the monolithic neural network**

The images obtained in the edge detection phase are used as the inputs of the neural network. In order to evaluate more objectively the recognition rate, the k-fold cross validation method was used. The training process is defined as follow.

1. Define the parameters for the monolithic neural network [15].

   - Layers hidden: two
   - Neuron number in each layer: 200
   - Learning algorithm: Gradient descent with momentum and adaptive learning.
   - Error goal: 1e-4.

2. The indices for training and test k folds were calculated as follow.

   - Define the people number (p).
   - Define the sample number for each person (s).

- Define the k-folds $(k = 5)$.
- Calculate the number of samples (m) in each fold by using (16)

$$m = (s/k) \cdot p \qquad (16)$$

- The train data set size (i) is calculated in (17)

$$i = m(k - 1) \qquad (17)$$

- Finally, the test data set size (18), are the samples number in only one fold.

$$t = m \qquad (18)$$

- The train set and test set obtained for the three face database used in this work are show in Table 1.

3. The neural network was training k-1 times, one for each training fold calculated previously.

4. The neural network was testing k times, one for each fold test set calculated previously.

Table 1 Information for the tested database of faces

Database	People number (p)	Samples number (s)	Fold size (m)	Training set size (i)	Test size (t)
ORL	40	10	80	320	80
Cropped Yale	38	10	76	304	76

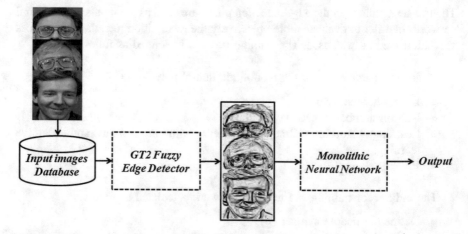

Fig. 5 General structure for the face recognition system

Finally, the mean of the rates of all the k-folds are calculated to obtain the recognition rate (Fig. 5).

6 Experimental Results

This section provides a comparison of the recognition rates achieved by the face recognition system when different fuzzy edge detectors were applied.

In the experimental results several edge detectors were analyzed in which are included the Sobel operator, Sobel combined with T1 FLS, IT2 FLS and GT2 FLS. Besides these, the Prewitt operator, Prewitt based on T1 FLS, IT2 FLS and GT2 FLS are also considered. Additional to this, the experiments were also validated without using any edge detector.

The tests were executed using the ORL and the Cropped Yale database; an example of these faces database is shown in Table 2. The parameters used in the monolithic neural network are described in Sect. 5. Otherwise, the training set and testing set that we considered in the tests are presented in Table 1, and these values depend on the database size used.

It is important to mention that all values presented below are the results of the average of 30 simulations achieved by the monolithic neural network. For this reason, the results presented in this section cannot be compared directly with the results achieved in [15]; because, in [15] only are presented the best solutions.

In the first test, the face recognition system was performed using the Prewitt and Sobel GT2 fuzzy edge detectors. This test was applied over the ORL data set. The

Table 2 Faces database

Database	Examples
ORL	
Cropped Yale	

Table 3 Recognition rate for ORL database using GT2 fuzzy edge detector

Fuzzy system edge detector	Mean rate (%)	Standard deviation	Max rate (%)
Sobel + GT2 FLS	87.97	0.0519	96.50
Prewitt + GT2 FLS	87.68	0.0470	96.25

mean rate, standard deviation and max rate values achieved by the system are shown in Table 3. In this Table we can note that better results were obtained when the Sobel GT2 fuzzy edge detector was applied; with a mean rate of 87.97, and standard deviation of 0.0519 and maximum rate of 96.50.

As a part of this test in Table 4, the results of 30 simulations are shown; these results were achieved by the system when the Prewitt GT2 fuzzy edge detector is applied.

Table 4 Recognition rate for ORL database using Prewitt GT2 fuzzy edge detector

Simulation number	Mean rate (%)	Standard deviation	Max rate (%)
1	83.50	0.0408	88.75
2	87.75	0.0357	92.50
3	89.75	0.0323	93.75
4	88.50	0.0427	95.00
5	70.25	0.3728	91.25
6	86.75	0.0420	91.25
7	89.00	0.0162	91.25
8	87.25	0.0205	88.75
9	86.75	0.0227	90.00
10	90.00	0.0441	**96.25**
11	89.50	0.0381	95.00
12	90.00	0.0265	92.50
13	89.75	0.0205	92.50
14	89.25	0.0360	92.50
15	92.00	0.0189	93.75
16	88.00	0.0447	93.75
17	86.25	0.0605	**96.25**
18	90.25	0.0323	95.00
19	90.50	0.0189	92.50
20	88.25	0.0227	91.25
21	89.00	0.0323	92.50
22	89.25	0.0167	91.25
23	89.00	0.0503	95.00
24	87.00	0.0447	91.25
25	89.25	0.0189	92.50
26	85.00	0.0776	93.75
27	87.00	0.0512	93.75
28	89.75	0.0399	95.00
29	86.00	0.0408	90.00
30	86.00	0.0503	93.75
Average	87.68	0.0470	92.75

In another test the system was considered using the Cropped Yale database. The numeric results for this experiment are presented in Table 5. In this Table we can notice that both edge detectors achieved the same max rate value; but, the mean rate was better with the Sobel + GT2 FLS.

As part of the goals of this work, the recognition rate values achieved by the system when the GT2 fuzzy edge detector is used, were compared with the results obtained when the neural network is training without edge detection, the Prewitt operator, the Prewitt combined with T1 and IT2 FSs; also, the Sobel operator, the Sobel edge detector combined with T1 and IT2 FSs. The results achieved after to apply these different edge detection methods are show in Tables 6 and 7.

Table 5 Recognition rate for Cropped Yale database using GT2 fuzzy edge detector

Fuzzy system edge detector	Mean rate (%)	Standard deviation	Max rate
Sobel + GT2 FLS	93.16	0.0328	100
Prewitt + GT2 FLS	97.58	0.0328	100

Table 6 Recognition rate for ORL database

Fuzzy system edge detector	Mean rate (%)	Standard deviation	Max rate
None	2.59	0.0022	5.00
Sobel operator	2.70	0.0037	5.00
Sobel + T1FLS	86.16	0.0486	93.75
Sobel + IT2FLS	87.35	0.0373	95.00
Sobel + GT2 FLS	**87.97**	**0.0519**	**96.50**
Prewitt operator	2.70	0.0036	5.00
Prewitt + T1FLS	87.03	0.0386	93.75
Prewitt + IT2FLS	87.54	0.0394	95.00
Prewitt + GT2 FLS	87.68	0.0470	96.25

Table 7 Recognition rate for Cropped Yale database

Fuzzy system edge detector	Mean rate (%)	Standard deviation	Max rate
None	2.83	0.0042	6.57
Sobel operator	2.63	0.0025	2.63
Sobel + T1FLS	97.52	0.0293	100
Sobel + IT2FLS	97.70	0.0314	100
Sobel + GT2 FLS	**98.11**	**0.0314**	**100**
Prewitt operator	2.80	0.0050	5.26
Prewitt + T1FLS	94.28	0.0348	100
Prewitt + IT2FLS	94.35	0.0304	100
Prewitt + GT2 FLS	97.58	0.0328	100

The results obtained for the ORL database are presented in Table 6; so, in this Table we can notice that the mean rate value is better when the Sobel GT2 fuzzy edge detector is applied with a value of 87.97. In these results we can also observe that the mean rate and max rate values obtained with the Prewitt + GT2 FLS were better than the Prewitt + IT2 FLS and Prewitt + T1 FLS.

Otherwise, the results achieved when the Cropped Yale database is used are shown in Table 7. In this Table we observed that the best performance (mean rate) of the neural network is obtained when the Sobel + GT2 FLS was applied; nevertheless, we can notice than the max rate values obtained by all the fuzzy edge detectors was of 100%.

7 Conclusions

In summary, in this paper we have presented two edge detector methods based on GT2 FS. The edge detection was applied in two image databases before the training phase of the monolithic neural network.

Based on the simulation results presented in Tables 6 and 7 we can conclude that the edge detection based on GT2 FS represent a good way to improve the performance in a face recognition system.

In general, the results achieved in the simulations were better when the fuzzy edge detection was applied; since the results were very low when the monolithic neural network was performed without edge detection; even so, when the traditional Prewitt and Sobel edge detectors were applied.

References

1. J. Canny, A computational approach to edge detection. IEEE Trans. Pattern Anal. Mach. Intell. **8**(6), 679–698 (1986)
2. I. Sobel, Camera models and perception. Ph.D. thesis, Stanford University, Stanford, CA, 1970
3. J.M.S. Prewitt, Object enhancement and extraction, ed. by B.S. Lipkin, A. Rosenfeld. in *Picture Analysis and Psychopictorics*, (Academic Press, NY, 1970), pp. 75–149
4. F. Perez-Ornelas, O. Mendoza, P. Melin, J.R. Castro, A. Rodriguez-Diaz, Fuzzy index to evaluate edge detection in digital images. PLoS ONE **10**(6), 1–19 (2015)
5. R. Kirsch, Computer determination of the constituent structure of biological images. Comput. Biomed. Res. **4**, 315–328 (1971)
6. L. Hu, H.D. Cheng, M. Zhang, A high performance edge detector based on fuzzy inference rules. Inf. Sci. **177**(21), 4768–4784 (2007)
7. Z. Talai, A. Talai, A fast edge detection using fuzzy rules, in *2011 International Conference on Communications, Computing and Control Applications (CCCA)*, Mar 2011, pp. 1–5
8. C. Tao, W. Thompson, J. Taur, A fuzzy if-then approach to edge detection, in *Fuzzy Systems*, (1993), pp. 1356–1360
9. R. Biswas, J. Sil, An improved canny edge detection algorithm based on type-2 fuzzy sets. Procedia Technol. **4**, 820–824 (2012)

10. O. Mendoza, P. Melin, O. Castillo, An improved method for edge detection based on interval type-2 fuzzy logic. Expert Syst. Appl. **37**(12), 8527–8535 (2010)
11. O. Mendoza, P. Melin, G. Licea, A new method for edge detection in image processing using interval type-2 fuzzy logic, in *2007 IEEE International Conference on Granular Computing (GRC 2007)*, Nov 2007, pp. 151–151
12. O. Mendoza, P. Melin, G. Licea, Interval type-2 fuzzy logic for edges detection in digital images. Int. J. Intell. Syst. (IJIS) **24**(11), 1115–1133 (2009)
13. C.I. Gonzalez, P. Melin, J.R. Castro, O. Mendoza, O. Castillo, An improved sobel edge detection method based on generalized type-2 fuzzy logic. Soft. Comput. **20**(2), 773–784 (2014)
14. P. Melin, C.I. Gonzalez, J.R. Castro, O. Mendoza, O. Castillo, Edge-detection method for image processing based on generalized type-2 fuzzy logic. IEEE Trans. Fuzzy Syst. **22**(6), 1515–1525 (2014)
15. O. Mendoza, P. Melin, O. Castillo, Neural networks recognition rate as index to compare the performance of fuzzy edge detectors, in *Neural Networks (IJCNN), The 2010 International Joint Conference on*, (2010), pp. 1–6
16. A. Doostparast Torshizi, M.H. Fazel Zarandi, Alpha-plane based automatic general type-2 fuzzy clustering based on simulated annealing meta-heuristic algorithm for analyzing gene expression data. Comput. Biol. Med. **64**, 347–359 (2015)
17. S.M.M. Golsefid, F. Zarandi, I.B. Turksen, Multi-central general type-2 fuzzy clustering approach for pattern recognitions. Inf. Sci. (Ny) **328**, 172–188 (2016)
18. G.E. Martínez, O. Mendoza, J.R. Castro, P. Melin, O. Castillo, Generalized type-2 fuzzy logic in response integration of modular neural networks, in *IFSA World Congress and NAFIPS Annual Meeting (IFSA/NAFIPS)*, (2013), pp. 1331–1336
19. J.M. Mendel, General type-2 fuzzy logic systems made simple: a tutorial. IEEE Trans. Fuzzy Syst. **22**(5), 1162–1182 (2014)
20. M.A. Sanchez, O. Castillo, J.R. Castro, Generalized type-2 fuzzy systems for controlling a mobile robot and a performance comparison with interval type-2 and type-1 fuzzy systems. Expert Syst. Appl. **42**(14), 5904–5914 (2015)
21. C. Wagner, H. Hagras, Toward general type-2 fuzzy logic systems based on zSlices. IEEE Trans. Fuzzy Syst. **18**(4), 637–660 (2010)
22. J.M. Mendel, R.I.B. John, Type-2 fuzzy sets made simple. IEEE Trans. Fuzzy Syst. **10**(2), 117–127 (2002)
23. D. Zhai, J.M. Mendel, Uncertainty measures for general type-2 fuzzy sets. Inf. Sci. **181**(3), 503–518 (2011)
24. D. Zhai, J. Mendel, Centroid of a general type-2 fuzzy set computed by means of the centroid-flow algorithm, in *Fuzzy Systems (FUZZ), 2010 IEEE International Conference on*, (2010), pp. 1–8
25. L.A. Zadeh, Outline of a new approach to the analysis of complex systems and decision processes. IEEE Trans. Syst. Man Cybern. **SMC-3**(1), 28–44 (1973)
26. L.A. Zadeh, *Fuzzy Sets*, vol. 8 (Academic Press Inc., USA, 1965)
27. F. Liu, An efficient centroid type-reduction strategy for general type-2 fuzzy logic system. Inf. Sci. **178**(9), 2224–2236 (2008)
28. X. Liu, J.M. Mendel, D. Wu, Study on enhanced Karnik-Mendel algorithms: Initialization explanations and computation improvements. Inf. Sci. **184**(1), 75–91 (2012)
29. J.M. Mendel, On KM algorithms for solving type-2 fuzzy set problems. IEEE Trans. Fuzzy Syst. **21**(3), 426–446 (2013)
30. C. Wagner, H. Hagras, Employing zSlices based general type-2 fuzzy sets to model multi level agreement, in *2011 IEEE Symposium on Advances in Type-2 Fuzzy Logic Systems (T2FUZZ)*, (2011), pp. 50–57
31. J.M. Mendel, Comments on α-plane representation for type-2 fuzzy sets: theory and applications. IEEE Trans. Fuzzy Syst. **18**(1), 229–230 (2010)
32. J.M. Mendel, F. Liu, D. Zhai, α-Plane representation for type-2 fuzzy sets: theory and applications. IEEE Trans. Fuzzy Syst. **17**(5), 1189–1207 (2009)

33. R.C. Gonzalez, R.E. Woods, S.L. Eddins, *Digital Image Processing using Matlab*, (Prentice-Hall, 2004)
34. O. Mendoza, P. Melin, G. Licea, A hybrid approach for image recognition combining type-2 fuzzy logic, modular neural networks and the Sugeno integral. Inf. Sci. **179**(13), 2078–2101 (2009)
35. The USC-SIPI image database, http://www.sipi.usc.edu/database/
36. A.S. Georghiades, P.N. Belhumeur, D.J. Kriegman, From few to many: illumination cone models for face recognition under variable lighting and pose. IEEE Trans. Pattern Anal. Mach. Intell. **23**(6), 643–660 (2001)
37. K.C. Lee, J. Ho, D. Kriegman, Acquiring linear subspaces for face recognition under variable lighting. IEEE Trans. Pattern Anal. Mach. Intell. **27**(5), 684–698 (2005)
38. P.J. Phillips, H. Moon, S.A. Rizvi, P.J. Rauss, The FERET evaluation methodology for face-recognition algorithms. IEEE Trans. Pattern Anal. Mach. Intell. **22**(10), 1090–1104 (2000)

Type-2 Fuzzy Logic Control in Computer Games

Atakan Sahin and Tufan Kumbasar

Abstract In this chapter, we will present the novel applications of the Interval Type-2 (IT2) Fuzzy Logic Controllers (FLCs) into the research area of computer games. In this context, we will handle two popular computer games called Flappy Bird and Lunar Lander. From a control engineering point of view, the game Flappy Bird can be seen as a classical obstacle avoidance while Lunar Lander as a position control problem. Both games inherent high level of uncertainties and randomness which are the main challenges of the game for the player. Thus, these two games can be seen as challenging testbeds for benchmarking IT2-FLCs as they provide dynamic and competitive elements that are similar to real-world control engineering problems. As the game player can be considered as the main controller in a feedback loop, we will construct an intelligent control systems composed of three main subsystems: reference generator, the main controller, and game dynamics. In this chapter, we will design and then employ an IT2-FLC as the main controller in a feedback loop such that to have a satisfactory game performance while be able to handle the various uncertainties of the games. In this context, we will briefly present the general structure and the design methods of two IT2-FLCs which are the Single Input and the Double Input IT2-FLCs. We will show that the IT2-FLC structure is capable to handle the uncertainties caused by the nature of the games by presenting both simulations and real-time game results in comparison with its Type-1 and conventional counterparts. We believe that the presented design methodology and results will provide a bridge for a wider deployment of Type-2 fuzzy logic in the area of the computer games.

A. Sahin (✉)
Centre for Process Analytics and Control Technology,
University of Strathclyde, Glasgow G1 1XL, UK
e-mail: atakan.sahin@strath.ac.uk

T. Kumbasar
Control and Automation Engineering Department,
Istanbul Technical University, 34469 Istanbul, Turkey
e-mail: kumbasart@itu.edu.tr

© Springer International Publishing AG 2018
R. John et al. (eds.), *Type-2 Fuzzy Logic and Systems*,
Studies in Fuzziness and Soft Computing 362,
https://doi.org/10.1007/978-3-319-72892-6_6

Keywords Type-2 fuzzy logic · Type-2 fuzzy sets · Type-2 fuzzy logic controllers · Computer games · Games · Lunar Lander · Flappy Bird

1 Introduction

Computer game industry is one of the biggest high-tech industry as well as its revenue. Depending on their virtual worlds which are inspired from real-world dynamics or facts, they are a perfect test-bed for computational intelligence methods or several types of research [1]. These research areas show extremely diversity depending on their field. For instance, self-awareness of the game bots is the challenging application of computational intelligence under computer science area [2–4]. On the opposite side, engineers try to design a perfect player to bear against game environment [5–7]. Furthermore, collected behavior logs of the human players from their plays might also be a source for social scientists [8]. Consequently, several games have been used as test beds such as Pacman [5], Scrabble [6], Super Mario [9], Counter-Strike [2], StarCraft [10], Flappy Bird [11, 12] and, Lunar Lander [13, 14].

In the last decade, Type-2 (T2) Fuzzy Logic, which is a generalization of ordinary (Type-1) fuzzy logic, has made a significant breakthrough in the area of computational intelligence [15, 16]. Especially, Interval T2 (IT2) Fuzzy Logic Controllers (IT2-FLCs) have been successfully employed in various engineering problems [17–23]. IT2-FLCs have the capability of handling high-level uncertainties as well as nonlinear dynamics in comparison with its Type-1 (T1) and conventional counterparts. This lies due to the extra degree of freedom provided by their T2 fuzzy sets (T2-FSs) [24, 25]. The superiority of IT2-FLCs has been shown in various control engineering applications such as in mobile robots [17–20], unmanned flight systems [21], engine control [22]. Although the mainstream of these researches are based on Double Input IT2-FLCs (DIT2-FLCs) [14, 16, 18–21, 23, 26], it has also been shown in [21, 27, 28] that Single Input IT2-FLCs (SIT2-FLCs) are easy to design and to deploy to real-time control engineering applications.

In this chapter, we will present the novel applications of the IT2-FLCs into the research area of computer games. In this context, we will handle two well-known computer games, namely Flappy Bird [11, 12] and Lunar Lander [13, 14], to show the abilities of the IT2-FLCs. From a control engineering point of view, as the game player can be seen as the main controller in a feedback loop, we will transform the game logic of flappy bird into a reference tracking problem while handling the moon landing problem as a position control problem. Thus, we will construct an intelligent control system composed of three main subsystems: reference generator, the main controller, and game dynamics. In this chapter, we will design and then employ an IT2-FLC as the main controller in a feedback loop such that to have a satisfactory performance and to be able to handle the various uncertainties of the games. In this context, we will briefly present the general structure and the design methods of two IT2-FLCs which are the SIT2-FLC and DIT2-FLC. In this chapter, we will design a SIT2-FLC for the game Flappy Bird while a DIT2-FLC structure for the game moon lander. The IT2-FLCs have been designed and implemented by using the Interval Type-2 Fuzzy Logic Toolbox [29] for Matlab/Simulink. We will examine the performance of both IT2 fuzzy control systems with respect to their control system and game performances, in comparison with its T1 and conventional counterparts, to show that the presented structure can handle the uncertainties caused by the nature of the games much better.

2 Interval Type-2 Fuzzy Logic Controllers

The aim of this section is to present the general structure of the PID type IT2-FLCs by classifying them on the number of input variables. In literature, the most widely used IT2-FLC structures are the SIT2-FLC and DIT2-FLC ones as presented in Fig. 1 [26, 28].

The SIT2-FLC structure uses only the error signal (e) as its input and the control signal (u_{SIT2}) as its output as shown in Fig. 1a [28]. The handled SIT2-FLC structure has one input and four output Scaling Factors (SFs). Here, K_E is the input SF that normalize the input to the common interval $[-1, +1]$ in which the membership functions (MFs) of the inputs are defined and is defined as:

$$K_E = \frac{1}{e_{max}} \tag{1}$$

where e_{max} represent the maximum allowed error value. While the IT2 fuzzy output (U) is mapped into the respective actual output (u_{SIT2}) by the output SFs as follows:

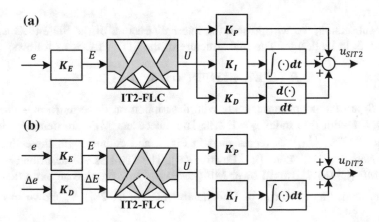

Fig. 1 Illustration of the **a** SIT2-FLC, **b** DIT2-FLC structure

$$u_{SIT2} = K_P U + K_I \int U dt + K_D \frac{dU}{dt} \tag{2}$$

where K_p, K_I and K_D are defined as:

$$K_P = K_{P0} K_u \quad K_D = K_{D0} K_U \quad K_I = K_{I0} K_u \tag{3}$$

Here, the output SF K_U is set as K_E^{-1} to rescale the IT2-FIS output while K_{P0}, K_{I0} and K_{D0} are the baseline PID controller gains.

The PID type DIT2-FLC is formed using PD type DIT2-FLC with an integrator and a summation unit at the output as shown in Fig. 1b [23, 26]. The PID type DIT2-FLC given in Fig. 1b is constructed by choosing the inputs to be an error (e) and change of the error (Δe) and the output as the control signal (u). Here, the input SFs K_E (for e) and K_d (for Δe) normalize the inputs to the universe of discourse where the MFs of the inputs are defined. Thus, e and Δe are converted after normalization into E and ΔE while the output (U) of the PID type DIT2-FLC is converted into the control signal (u) by the output SFs K_P (proportional SF) and K_I (integral SF) as follows:

$$u_{DIT2} = K_P U + K_I \int U dt \tag{4}$$

It can be concluded that both the SIT2-FLC and DIT2-FLC controllers are analogous to the conventional PID controllers from the input-output relationship point of view [23, 26, 28]. The main difference of these structure lies in the characteristics of their internal structure.

2.1 Single Input Interval Type-2 Fuzzy Logic Controller

In this subsection, we will present the internal structure of the SIT2-FLC and the key factors in its design. The rule structure of the SIT2-FLC is as follows:

$$R_q: \text{IF } E \text{ is } \tilde{A}_{1j} \text{ THEN } U \text{ is } B_j \quad j = 1, 2, 3 \tag{5}$$

where B_j are the crisp consequents with description on these as $B_1 = -1$, $B_2 = 0$ and, $B_3 = 1$ with rule index $q = 1, 2, 3$. The antecedent MFs are defined with triangular IT2-FSs $(\tilde{A}_{1j})$ as represented in Fig. 2 and defined with three linguistic terms: Negative (N), Zero (Z), Positive (P). The IT2-FSs are described with an upper MF (UMF) $\bar{\mu}_{\tilde{A}_{1j}}$ and Lower MF (LMF) $\underline{\mu}_{\tilde{A}_{1j}}$ that provide an extra degree of freedom named as Footprint of Uncertainty (FOU) [26, 28]. As shown in Fig. 2,

Fig. 2 Illustration of the antecedent of the IT2-FS

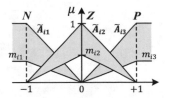

m_{ij}'s represent the height of the LMFs and will be the main design parameters of the SIT2-FLC to be tuned. For the sake of simplicity, we employ $m_{i2} = \alpha$ and $m_{i1} = m_{i3} = 1 - \alpha$. Thus, α is the new design parameter which determines the FOU in the antecedent IT2-FSs [28].

The implemented the SIT2-FLC uses the center of sets type reduction method [21, 27, 28]. It has been demonstrated in [28] that the defuzzified output can be calculated as:

$$U = \frac{U_l + U_r}{2} \tag{6}$$

where U_l and U_r are the left and right end points respectively of the type reduced set, are defined as follows:

$$U_l = \frac{\sum_{q=1}^{R} \underline{\mu}_{\tilde{A}1q} B_q + \sum_{q=R+1}^{N} \bar{\mu}_{\tilde{A}_{1q}} B_q}{\sum_{q=1}^{R} \underline{\mu}_{\tilde{A}_{1q}} + \sum_{q=R+1}^{N} \bar{\mu}_{\tilde{A}_{1q}}} \tag{7}$$

$$U_r = \frac{\sum_{q=1}^{L} \bar{\mu}_{\tilde{A}_{1q}} B_q + \sum_{q=L+1}^{N} \underline{\mu}_{\tilde{A}_{1q}} B_q}{\sum_{q=1}^{L} \bar{\mu}_{\tilde{A}_{1q}} + \sum_{q=L+1}^{N} \underline{\mu}_{\tilde{A}_{1q}}} \tag{8}$$

where R and L are the switching points [28]. As shown in Fig. 2, the SIT2-FLC employs fully overlapping IT2-FSs in the sense of LMFs and UMFs. Hence, it is guaranteed that a crisp value E' always belongs to two successive IT2-FSs $(\tilde{A}_{1j} \& \tilde{A}_{1j+1})$. Thus, since only $N = 2$ rules will be always activated, the values of R and L are equal to 1 [28]. Moreover, the input-output mapping of the SIT2-FLC $(U(E))$ can be derived as follows [27, 28]:

$$U(E) = E \cdot k(|E|) \tag{9}$$

where, $k(E)$ is the T2 fuzzy gain and defines as:

$$k(E) = \frac{1}{2} \left(\frac{1}{\alpha + E - \alpha E} + \frac{\alpha - 1}{\alpha E - 1} \right) \tag{10}$$

Accordingly, the main design parameter of the IT2 FSs, α, is assigned as the only tuning parameter. This derivation simplifies the SIT2-FLC design into a

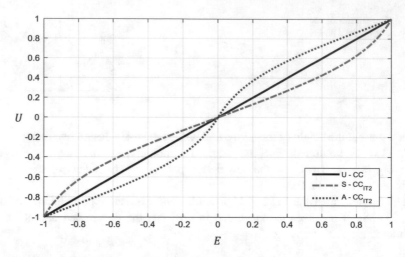

Fig. 3 Illustration of the $CC_{IT2}s$

Control Curve (CC) generation instead of conventional control surface design. In [28], by defining $\varepsilon_0(E) = U(E) - E$, the following design guidelines have been presented.

- If $0 < \alpha \leq \alpha_{c1}$, then $\varepsilon_0 < 0$ for $\forall E \in O_S$ where $O_S \in [0, 1)$ and $\alpha_{c1} = (3 - \sqrt{5})/2$. Thus, a Smooth $CC_{IT2}(S - CC_{IT2})$ will be generated.
- If $\alpha_{c2} \leq \alpha < 1$, then $\varepsilon_0 > 0$ for $\forall E \in O_A$ where $O_A \in [0, 1)$ and $\alpha_{c2} = (\sqrt{5} - 1)/2$. Thus, an Aggressive CC_{IT2} $(A - CC_{IT2})$ will be generated.

In Fig. 3, a $S - CC_{IT2}$ and $A - CC_{IT2}$ examples are given for $\alpha = 0.2$ and $\alpha = 0.8$, respectively. Here, a Unit CC (U – CC) is sketched for the comparison. It can be clearly seen that the $S - CC_{IT2}$ has relatively low input sensitivity when E is close to "0" when compared to the $A - CC_{IT2}$. Thus, the parameter α of the SIT2-FLC can be tuned such that to enhance the control performance of its baseline counterpart via the design guidelines [28].

2.2 Double Input Interval Type-2 Fuzzy Controller

In this subsection, we will present the internal structure of the DIT2-FLCs. The DIT2-FLC, handled in this study, uses and employs the 3×3 rule base given in Table 1. The rule structure is defined as follows:

Table 1 Rule table

$\Delta E/E$	N	Z	PB
N	NB	N	Z
Z	N	Z	P
P	Z	P	PB

$$R_q: \text{IF } E \text{ is } \tilde{A}_{1i} \text{ and } \Delta E \text{ is } \tilde{A}_{2j} \text{ THEN } U \text{ is } C_q \tag{11}$$

where C_q is the crisp consequent MFs $(q = 1, \ldots, Q = 9)$ is defined five linguistic terms Negative Big (NB), N, Z, P and Positive Big (PB) that represent $-1, -0.5, 0, 0.5, 1$, respectively. The antecedent part of the rule is defined with IT2-FSs $(\tilde{A}_{1i}, \tilde{A}_{2j}; i = 1, 2, 3; j = 1, 2, 3)$ which are defined with three linguistic terms N, Z and P. The IT2-FSs can be described with UMFs $\left(\bar{\mu}_{\tilde{A}_{1i}} \text{ and } \bar{\mu}_{\tilde{A}_{2j}} \right)$ and LMFs $\left(\underline{\mu}_{\tilde{A}_{1i}} \text{ and } \underline{\mu}_{\tilde{A}_{2j}} \right)$ which provides extra degree of freedom that is also known as FOU. Similar to its input counterpart, the FOU of IT2-FSs is generated with the heights of the LMFs (m_{ij}) which is the only design parameter to be tuned [23, 26].

The implemented DIT2-FLC uses the center of the sets type reduction method [26]. It has been demonstrated in [26] that the defuzzified output can be calculated as:

$$U = \frac{U_l + U_r}{2} \tag{12}$$

where U_l and U_r are the left and right end points respectively of the type reduced set, are defined as follows:

$$U_l = \frac{\sum_{q=1}^{L} \bar{f}_q C_q + \sum_{q=L+1}^{Q=9} \underline{f}_q C_q}{\sum_{q=1}^{L} \bar{f}_q + \sum_{q=L+1}^{Q=9} \underline{f}_q} \tag{13}$$

$$U_r = \frac{\sum_{q=1}^{R} \underline{f}_q C_q + \sum_{q=R+1}^{Q=9} \bar{f}_q C_q}{\sum_{q=1}^{R} \underline{f}_q + \sum_{q=R+1}^{Q=9} \bar{f}_q} \tag{14}$$

where R and L are the switching points defined between $[1, Q-1]$ [15–17]. Moreover, $\tilde{f}_q = \begin{bmatrix} \bar{f}_q & \underline{f}_q \end{bmatrix}$ is the total firing strength for the qth rule and is defined as:

$$\underline{f}_q = \underline{\mu}_{\tilde{A}_{1i}} * \underline{\mu}_{\tilde{A}_{2j}} \tag{15}$$

$$\bar{f}_q = \bar{\mu}_{\tilde{A}_{1i}} * \bar{\mu}_{\tilde{A}_{2j}} \tag{16}$$

Here, "*" represents the product implication (the t-norm). The typed reduced set can be calculated by finding the optimal switching points (R and L) with the Karnik and Mendel algorithm [15, 16, 26].

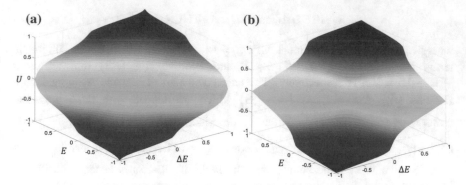

Fig. 4 Illustration of the **a** smooth, **b** aggressive control surface of DIT2-FLCs

In control engineering applications, it is usually desired to design a symmetric control surface [20, 26, 28, 30]. In [31, 32], it has been shown that, by setting $m_{11} = m_{13}$ and $m_{21} = m_{23}$, how to generate smooth and aggressive control surfaces by simply tuning the FOU parameters. In Fig. 4a, a smooth control surface is presented for the FOU parameter settings $m_{11} = m_{13} = 0.3$, $m_{12} = 0.9$, $m_{21} = m_{23} = 0.3$ and $m_{22} = 0.9$ while an aggressive control surface is presented in Fig. 4b for the FOU parameter settings $m_{11} = m_{13} = 0.9$, $m_{12} = 0.1$, $m_{21} = m_{23} = 0.9$ and $m_{22} = 0.1$.

3 Type-2 Fuzzified Flappy Bird Control System

In this section, we will represent the design and performance evaluation of the T2 Fuzzified Flappy Bird Control System for the game Flappy Bird. Flappy Bird is published as a mobile game in 2013 by GEARS Studios [33]. As a side-scrolled game, the basic game logic of the game is as follows: each time the player taps the screen, the bird flaps its wings, moving upward in an arc while gravity constantly pulls downward; if the screen is not tapped, the bird falls to the ground due to gravity, which also ends the game. The main goal of the game is to control the bird's height while attempting to fly between the obstacles (i.e. the pipes) without hitting them [11, 33].

3.1 Game Space

Here, we will briefly explain the game space of the Matlab replica of the game Flappy Bird that can be downloaded from [34]. In this version of this game, the game parameters are grouped as the parameters of the environment

Fig. 5 Illustration of the game space and one of the sample generated reference for Flappy Bird

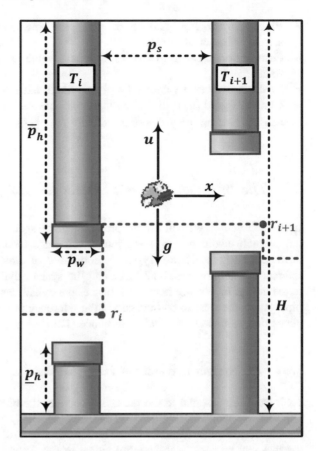

$\left(p_g, \bar{p}_h, \underline{p}_h, p_s, p_w, H \right)$ and the bird's dynamics (g, v, u). These parameters are defined as follows with their illustration in Fig. 5 [11, 33]:

- **World height** (H) is the distance between the ceiling and the floor with a fixed value of 180 pixels.
- **Upper pipe height** $(\bar{p}_h)$ is the distance between the top of the upper pipe and ceiling. This value is generated by a uniformly distributed random number generator.
- **Pipe gap** (p_g) is the distance between the pipes which is fixed as 49 pixels.
- **Lower pipe height** $\left(\underline{p}_h \right)$ is the distance between the top of the lower pipe and floor. The value of the lower pipe height is defined as $\underline{p}_h = \bar{p}_h - 49$. Though, it is constrained with a minimum value as 36 pixels.
- **Pipe width** (p_w) is the width of the pipe with a fixed value of 24 pixels.

- **Pipe separation** (p_s) is the horizontal space between two consequent pipes with a fixed value of 80 pixels.
- **Gravitational constant** (g) has a default value assigned as 0.1356 pixels per frame.
- **Bird's x-direction velocity** (v_x) has a fixed value of 1 pixel per frame.
- **Control signal** (u) is the binary input variable $u \in \{0, 1\}$ provided by the user to flap the bird on the y direction velocity (v_y) [11].

3.2 The Intelligent Control System for Flappy Bird

In this subsection, we will use the presented game space of Flappy Bird and convert the obstacle avoidance problem into a reference tracking control problem. The proposed T2 fuzzy control system is composed of three main parts which are the reference generator, the SIT2-FLC, and the system dynamics of the bird as illustrated in Fig. 6. We will handle the bird as the dynamic system to be controlled, the pipe gap as the goal to be tracked via the reference trajectory and the environment generation as uncertainty and disturbance [12].

3.2.1 The System Dynamics of Flappy Bird

As it has been asserted in the preceding section, the bird has a constant horizontal (x) velocity while the vertical (y) velocity depends on the player's taps which directly controls the dynamics of the bird. From a control engineering point of view, the taps can be seen as the control signal of the system which is based on binary numbers and the vertical velocity to be controlled. In the rest of the section, we will use the abbreviation v for representing v_y since v_x is constant variable as described in the game logic. The bird's system dynamics are defined as follows [34]:

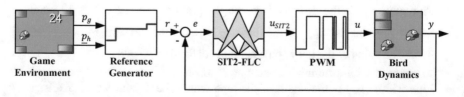

Fig. 6 Illustration of the T2 fuzzified Flappy Bird control system

$$y_t = y_{t-1} + v_t$$

$$v_t = \begin{cases} 2.5, & u = 1 \\ v_{t-1} - g, & u = 0 \end{cases} \tag{17}$$

where v_t and y_t are vertical velocity and vertical position of the bird at tth frame as respectively.

3.2.2 The Reference Generator

The reference generator is an essential component in the control loop as it transforms the obstacle avoidance problem of the game to a fuzzy feedback control system. The reference generator provides the trajectory for the bird by taking account the gap between the pipes and the bird's position. The reference trajectory (r_i) is updated when the bird reaches the end of the pipe set (T_i) automatically as shown in Fig. 5 (the red line). The new reference trajectory is defined as:

$$r_{i+1} = \underline{p_h} + 0.3\left(\bar{p}_h - \underline{p_h}\right) \tag{18}$$

where i is the frame when the bird reaches the end of the pipe.

3.2.3 The Interval Type-2 Fuzzy Logic Controller Structure

For the presented fuzzy control system in Fig. 6, we will prefer a SIT2-FLC structure given in Fig. 1a. Furthermore, as it is crucial not to hit and not to track the reference signal with zero steady state error, we will prefer to employ a P type SIT2-FLC for the sake of simplicity ($K_{I0} = 0, K_{D0} = 0$). In this structure, we will set and fix the input SF K_E as $K_E = 1/150$ while the output SF K_{P0} will be set and fixed to its baseline counterpart (its design will be explained in Sect. 3.3). Thus, the only parameter to be tuned in the SIT2-FLC structure is the FOU design parameter α. It is worth to note that the generated signal from the P type SIT2-FLC needs also to be converted to a binary signal. Thus, the continuous control signal (u_{IT2}) is then converted into a Pulse Width Modulation (PWM) generator, which is widely used in power electronics [35], into the input signal $u \in \{0, 1\}$.

3.3 The Design and Performance Evaluation of the SIT2-FLC Structure

This subsection will include the design steps of the SIT2-FLC and then investigate its performance in comparison with the conventional P controller. Then, we will

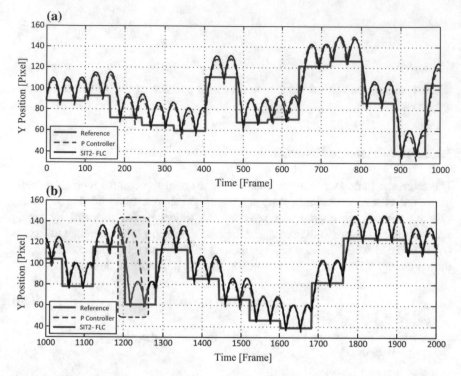

Fig. 7 System responses of the feedback control systems for the **a** training reference trajectory **b** testing reference trajectory

present experimental results that performed in both the simulation and game environment to examine its game performance.

As it has been asserted in Sect. 2.1, the design of the SIT2-FLC is accomplished as an extension of its baseline counterpart. Thus, as it has been preferred to design and employ a P type SIT2-FLC, it is necessary to design a conventional P controller. In this context, the Genetic Algorithm is used to find the optimal proportional gain value K_{P0} on a randomly generated training reference trajectory regarding the game logic as illustrated in Fig. 7a. The defined cost function is the widely used Integral of Time Absolute Error (ITSE) which is given below:

$$\text{ITSE} = \int e^2(t) t dt \tag{19}$$

The resulting optimal parameter value for the proportional gain is found as $K_{p0} = 5.04$ from the training trajectory. Then, to enhance the control performance of its baseline counterpart in presence of uncertainties and nonlinearities, we preferred to set the FOU design parameters value as $\alpha = 0.2$ to end up with $S - CC_{IT2}$ which is potentially more robust controller in comparison with its baseline and $A - CC_{IT2}$

Table 2 Control performance evaluation for Flappy Bird

	Average	Best
P controller	112	423
SIT2-FLC	174	482

counterparts. The control system performances of the SIT2-FLC and P controller structures are given in Fig. 7a, and their corresponding ITSE values are 30130 and 31630, respectively. It can be concluded that the SIT2-FLC reduced the ITSE value by about 5% in comparison to its baseline counterpart in the training phase. Moreover, since the dynamics of flappy bird system inherent nonlinearity as given in (17), we have examined the controller performances for a testing trajectory which is also generated randomly as shown in Fig. 7b. In other words, we have tested the controller performances for different operating points at which they have not been designed. The ITSE values of the SIT2-FLC and P controller structures are calculated as 99674 and 136270, respectively. Thus, in comparison with its conventional counterpart, the SIT2-FLC resulted with a better tracking performance as shown in Fig. 7b and was also able to reduce the ITSE value by about 27% on the testing trajectory. Moreover, with respect to the game logic; it is worth to underline at the reference variation (r_i) in the 1200th frame (the shaded area in Fig. 7b) that the conventional control system almost hit the T_{i+1} pipe which would end the game. On the other hand, for the same operation point, the T2 fuzzified flappy bird control system resulted with a satisfactory reference tracking performance.

In the real game environment, the ITSE value comparison loses its importance as the number of successfully avoided pipes is the indicator of the score rather than the reference tracking performance. Therefore, since the game environment parameters such as pipe gap's location $\left(\bar{p}_h, \underline{p}_h\right)$ are generated randomly during the game, we have repeated each experiment 20 times to get an overall performance comparison. The results of the game performances are tabulated in Table 2 where the best and average scores of the experiments are given. It can be clearly observed that the T2 fuzzy control scheme improved the average score almost by 55% in comparison with its baseline counterpart. Consequently, the SIT2-FLC structure is better when compared to its conventional counterpart with respect to both its control and game performance.

4 Type-2 Fuzzy Moon Landing System

In this section, we will represent the design and performance evaluation of the T2 Fuzzy Moon Landing System for the game Lunar Lander. Lunar Lander is one of the most cloned games [36] into several platforms which is a vector monitor based arcade game firstly released by Atari in the late 1970s. The game logic of the Lunar Lander game is to control the engine of the spaceship in the x-y coordinate system such that to land on the dock softly [37, 38]. The player can arrange the spaceship's angular rotation by pressing the right or left arrow keys on a cumulative basis. The player can

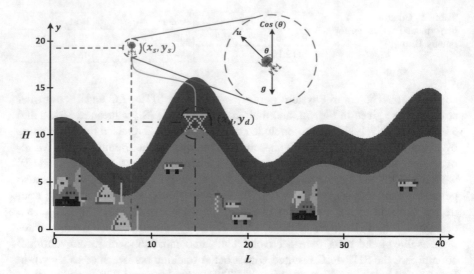

Fig. 8 Illustration of the game environment of Lunar Lander

also produce thrust against the gravity by pushing the spacebar key. The direction of the force obviously depends on the spaceship's angular position [12, 13].

4.1 Game Space

In this study, we will use the Matlab clone of the Lunar Lander that can be found in [39]. The game parameters of the game space are illustrated in Fig. 8 and defined as follows

- **World height** (H) is fixed to a value of 20 units.
- **World length** (L) is fixed value of 40 units.
- **The terrain** is defined and generated with randomly generated sinus functions. In Fig. 8, light gray basements define the basic terrain, which is also an obstacle to be avoided by the player. On the other hand, the dark gray ones only provide a realistic game environment, thus are not obstacles.
- **Position of the Dock** (x_d, y_d) is the reference point or the target position for landing on x-y coordinates. The x-axis is generated as randomly while the y-axis depends on the randomly generated light gray terrain.
- **Position of the Spaceship** (x_s, y_s) is the position of the spaceship on x-y coordinate. In the start of the game, the x-axis position of the spaceship is randomly assigned in the range of the world length (L) while its y-axis position is set to a 20 units.

- **Speed of the Spaceship** (v_x, v_y) defines the linear speed of spaceship in x-y the coordinate system.
- **Gravitational constant** (g) has a default value assigned as 0.4 units per frame.
- **The angular position of the Spaceship** (θ) is the angle between the spaceship's direction and y-axis. The angle can be controlled with right and left arrow keys by the player.
- **Thrust power** (T) defines the thrust force to be employed to the spaceship engine and is controlled by the player.
- **Fuel** $(\int T)$ defines the maximum total thrust power that can be consumed by the player.

In this Matlab clone [39], the game ends when

(1) The spaceships touches/hits the terrain with failure
(2) The spaceship consumes more than 200-unit fuel before a successful landing with failure
(3) The Euclidean distance between the spaceship and the dock is less than 2 units with successful landing

Moreover, we have added the following conditions for a successful landing to make the game more realistic with providing soft landing conditions:

(4) Vertical velocity (v_y) must be less than -0.5 when the spaceship has landed.
(5) The angular position of the spaceship (θ) must be between $[-\pi/16, \pi/16]$ when the spaceship has landed.

4.2 The Intelligent Control System for Lunar Lander

In this section, we will convert the defined game space of Lunar Lander into a feedback control problem. Here, the spaceship will define as the system to be controlled and the position of the dock (x_d, y_d) to be the desired reference point.

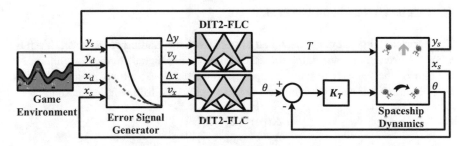

Fig. 9 Illustration of the type-2 fuzzy moon landing system

Randomly generated initial spaceship position and terrain, and the gravity will be considered as the disturbances and uncertainties in the control loop. Moreover, error signal generator will convert the position signals from the dock (x_d, y_d) and spaceship (x_s, y_s) with game disturbances such as g into feasible reference signal for the controllers that will define to control spaceship's angular position and thrust power via angle (θ) and thrust (T) signals. Therefore, we can define the proposed structure into three part as illustrated in Fig. 9. Note that, we have also designed an inner loop proportional controller to speed up the response time of angle of the spaceship (θ). In all experiments, we have set and fixed this controller gain as $K_T = 1.648$.

4.2.1 The System Dynamics

The dynamics of the spaceship are based on the classical motion equations. Thus, the acceleration (a) of the spaceship at kth frame in x-y coordinate system can be defined as:

$$
\begin{aligned}
a_x(k) &= -\sin(\theta(k))T(k)d_t \\
a_y(k) &= \cos(\theta(k))T(k)d_t
\end{aligned}
\tag{20}
$$

where d_t is sampling time of the game with a fixed value of 0.1. Correspondingly, the velocity (v) equations of the both axis can be defined as:

$$
\begin{aligned}
v_x(k+1) &= v_x(k) + a_x(k)d_t \\
v_y(k+1) &= v_y(k) + a_y(k)d_t - gd_t
\end{aligned}
\tag{21}
$$

Moreover, the position of the spaceship (x_s, y_s) can be defined as:

$$
\begin{aligned}
x_s(k+1) &= x_s(k) + v_x(k)d_t \\
y_s(k+1) &= y_s(k) + v_y(k)d_t
\end{aligned}
\tag{22}
$$

4.2.2 The Error Signal Generator

The error signal generator is designed to transform the landing of the spaceship into a control problem by providing the essential reference signals to the controllers. As the aim of the game is to land the spaceship on the dock, the position differences between the dock and spaceship are used to define the following error signals:

$$
\begin{aligned}
\Delta x(k) &= x_d - x_s(k) \\
\Delta y(k) &= y_d - y_s(k)
\end{aligned}
\tag{23}
$$

Note that, the values x_d and y_d are fixed to the randomly generated values at the beginning of the game until the game ends.

4.2.3 The Interval Type-2 Fuzzy Logic Controller Structure

As it has been asserted, the player has to control the thrust and angle of the spaceship for a successful landing. In this context, we will design two DIT2-FLCs to control the thrust power and angular position of the spaceship to provide a successful landing as illustrated in Fig. 9. Here, we preferred a PD type DIT2-FLC structure since according to the success criteria of the game velocity and position must be smaller than predefined values. Note that, we have not preferred a PID structure, as it is necessary to eliminate the steady state error according to the definitions of a successful landing presented in Sect. 4.1 with 3rd condition for the game ending. As it can be seen from Fig. 9, we employed two PD type DIT2-FLC to solve the thrust (T) and angle (θ) control problems of the lunar lander. The DIT2-FLC structure for angle control is constructed by choosing the inputs as $e = \Delta x(k)$, $\Delta e = v_x(k)$ and the output as $u = \theta(k)$. In a similar manner, the DIT2-FLC structure for thrust control is constructed with $e = \Delta y(k)$ and $\Delta e = v_y(k)$ and $u = T(k)$. For each DIT2-FLC, there are 2 SFs, excluding the SF K_E, and 6 FOU parameters to be tuned, thus in total 2×8 parameters for DIT2-FLC structure.

4.3 The Design and Performance Evaluation of the DIT2-FLC Structure

This subsection will include the design steps of the DIT2-FLC structure and investigate its performance in comparison with its T1 fuzzy and conventional PD counterparts. Then, to examine the differences on the different level of the uncertainties, experimental results collected from the game environment are presented. The T1-FLC structure is composed with the identical rules of the DIT2-FLC (Table 1) ones with the only difference that it uses and employs triangular type T1-FSs [25].

As it has been mentioned, the Lunar Lander is a limited type game depending on game ending condition and also includes random parameter initializations for each trial. Therefore, the parameter tuning phase of the controllers should be accomplished with several trials in the game space to design controllers that are robust for randomly generated game environments. To provide that, we defined 4 training sets, as tabulated in Table 3, and then tuned the controllers respectively. The starting position of the spaceship is set and fixed during this phase to the value (16, 20). Moreover, the terrain characteristics have also been set and fixed in the training phase to make a fair comparison between the controllers. Here, all three controller

Method	Controller	Parameter	Value
Table 3 Controller parameters for Lunar Lander			
PD	Thrust	K_P	5.22
		K_D	13.14
	Angle	K_P	0.51
		K_D	6.12
T1-FLC	Thrust	K_E	$1/e_{max}$
		K_P	39.51
		K_D	0.18
	Angle	K_E	$1/e_{max}$
		K_P	41.37
		K_D	0.61
DIT2-FLC	Thrust	m_{11}, m_{13}	0.27
		m_{12}	0.91
		m_{21}, m_{23}	0.11
		m_{22}	0.67
	Angle	m_{11}, m_{13}	0.15
		m_{12}	0.61
		m_{21}, m_{23}	0.47
		m_{22}	0.91

structures were optimized with the particle swarm optimization that subject to minimization of the given objective function:

$$F = \sum_{k=1}^{n} \|e_\Delta(k)\|^2 + C \qquad (24)$$

where n represents the total number of samples which has been taken for each sampling time starting from the beginning to landing or crash. $\| \cdot \|$ is norm operator for the error term $e_\Delta(k)$ which is defined as:

$$e_\Delta(k) = [\Delta x, \Delta y]_{(1 \times 2)} \qquad (25)$$

Moreover, C is the penalty for a crash defined as:

$$C = \begin{cases} 10000, & if\ crash = 1 \\ 0, & if\ crash = 0 \end{cases} \qquad (26)$$

Note that, to show the superiority of DIT2-FLCs clearly, we have not optimized the SFs of the DIT2-FLCs. We have set and fixed them to the optimal values found for its T1 counterpart. The resulting optimal parameters are tabulated in Table 3 according to the training scenarios tabulated in Table 4. To provide a further comparison, we have also provided the resulting Landing Times (LT) and the existence of a crash. Firstly, it can be clearly observed that all three control

Table 4 Control performance evaluation of training and testing scenarios for Lunar Lander

	Scenario	$x_s(0), y_s(0)$	x_d, y_d	PD			T1-FLC			DIT2-FLC		
				F	LT	$\mathcal{C}$	F	LT	$\mathcal{C}$	F	LT	$\mathcal{C}$
Training	1	(16, 20)	(2, 5, 4)	1612	17.6	N	1508	14.2	N	1520	14.6	N
	2		(10, 4.2)	1577	21.2	N	1278	16.7	N	1307	17.0	N
	3		(18, 5.6)	1284	19.9	N	1087	12.1	N	905	11.7	N
	4		(25, 4.3)	1626	20.7	N	1394	17.2	N	1372	16.5	N
Testing	5	(35, 20)	(2, 5, 4)	13093	–*	Y	3938	23.5	N	3775	22	N
	6		(5, 3.6)	12932	–*	Y	13107	–*	Y	3500	22.4	N
	7		(10, 4.2)	12394	–*	Y	12607	–*	Y	2793	20.4	N
	8		(35, 5.5)	1368	20.5	N	10915	–*	N	1317	20.6	N

*Not applicable because of the crash

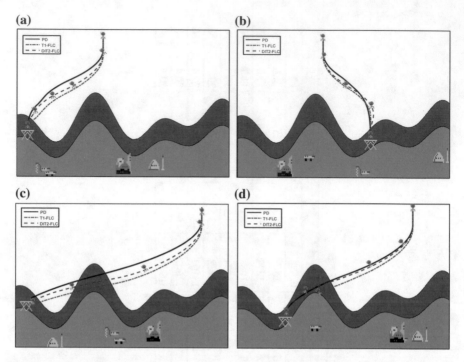

Fig. 10 Illustration of some training and testing cases. **a** Scenario-1, **b** Scenario-4 for training, **c** Scenario-5, **d** Scenario-7 for testing

structures achieved their training scenarios without a crash as expected from the training phase. The system responses for Scenario 1 and 4 are illustrated in Fig. 10a and Fig. 10b, respectively. For Scenario 3, the DIT2-FLC structure decreased the Landing Time value about 41 and 3% (increased the convergence speed to dock) while it also reduced the total fitness value about 30 and 16% in comparison to the PD and T1-FLC structures, respectively. It should be noted that the performances of the T1 and T2 Fuzzy Moon Landing Systems are quite similar. That lies because the design of the DIT2-FLC has been accomplished as an extension of its T1 counterpart. Moreover, this also coincides with the results presented in [20] where it has been stated that the DIT2-FLCs result in smoother control surfaces in comparison with its T1 counterpart. Thus, the resulting system response might be relatively slower but potentially more robust against uncertainties. Similar comments can be made for the other two reference variations.

We have also tested the controllers for a different initial point (35, 20) to see the performance of the controllers in terms of uncertainties and different operating regions. The resulting performance values are also tabulated in testing part of Table 4. The system responses for Scenario 5 and 7 are illustrated in Fig. 10c and Fig. 10d, respectively. It can clearly observe that the T2 fuzzy moon landing system was able to pilot the spaceship to the dock without a crash for all testing scenarios while T1 and conventional PD controller structures crashed the spaceship in three

Table 5 Game performance evaluation on 200 trials for Lunar Lander

	Crash count	Success rate (%)
PD	78	61
T1-FLC	61	69.5
DIT2-FLC	33	83.5

of them. The PD structure was not able to handle the uncertainty and thus, in the first three testing scenarios, the crash occurred for several reasons. Scenario 5 was failed because the 4th condition of the successful landing (presented in Sect. 4.1) was violated. PD structure has also violated 1st condition at Scenario 6 and 7. The T1 fuzzy structure crashed the spaceship in Scenario Numbers 6 and 7 since it hit/touched the terrain (1st condition). The last crash (Scenario 8) of the T1-FLC structure occurred due to the fact the required angle condition (5th condition) for a successful landing could not be satisfied as it resulted with oscillating system response. The handled scenarios clearly show that the proposed T2 fuzzy moon landing system can handle uncertainties and various operating points when compared to its T1 and conventional PD controller counterparts.

Distinctly from the simulation studies, the game environment includes various uncertainty sources and operating points caused by its randomization processes in the game logic such as various unique terrains, starting points, and landing points. Therefore, using squared error based evaluation criteria for comparison in the game environment might not be meaningful. Thus, we will use the successful landing criteria for the testing and compare the controllers in the game environment. In this context, we have employed 200 times each controller structure to game where starting point (x_s, y_s), landing point (x_d, y_d) and also terrain characteristics are randomly generated by the game. The success rates of the controllers are given in Table 5. It can be concluded that the game performance of the DIT2-FLC structure, with respect to game logic, is better than its T1 and conventional parts by almost 14% and 22.5%, respectively.

5 Conclusions

In this chapter, we presented the novel applications of the IT2-FLCs into the well-known computer games called Flappy Bird and Lunar Lander. As these games include various dynamics, and the uncertainties; these two games are challenging testbeds for benchmarking IT2-FLCs as they provide similar real-world engineering problems. From a control engineering point of view, as the game player can be seen as the main controller in a feedback loop, we have transformed the game logic of flappy bird into a reference tracking problem while the moon landing problem as a position control problem. Then, we proposed an intelligent control system where the IT2-FLC is the main controller. We designed a SIT2-FLC for the game Flappy Bird while a DIT2-FLC structure for the game moon lander. We examined the performance of both IT2 fuzzy control systems with respect to their control system and game performances in comparison with its T1 and conventional counterparts. Thereby, we have shown that the resulting IT2-FLCs resulted with an adequate control and game performance in the presence of the uncertainties, disturbances and nonlinear system dynamics.

References

1. S.M. Lucas, G. Kendall, Evolutionary computation and games. IEEE Comput. Intell. Mag. **1**(1), 10–18 (2006)
2. N. Cole, S.J. Louis, C. Miles, Using a genetic algorithm to tune first-person shooter bots, in *Proceedings of IEEE Congress on Evolutionary Computing* (2004), pp. 139–145
3. J. MacGlashan, M.L. Littman, Between imitation and intention learning, in *Proceedings of International Conference on Artificial Intelligence* (2015), pp. 3692–3698
4. P. Hingston, A turing test for computer game bots. IEEE Trans Comput Intell. AI Games **1**(3), 169–186 (2009)
5. S.M. Lucas, Evolving a neural network location evaluator to play ms. pac-man, in *Proceedings of IEEE Symposium on Computational Intelligence and Games* (2005), pp. 203–210
6. B. Sheppard, World-championship-caliber Scrabble. Artif. Intell. **134**(1), 241–275 (2002)
7. D. Perez, G. Recio, Y. Saez, P. Isasi, Evolving a fuzzy controller for a car racing competition, in *Proceedings of IEEE Symposium on Computational Intelligence and Games*, September 2009, pp. 263–270
8. W. Goufang, F. Zhou, L. Ping, L. Bo, Shaping in reinforcement learning via knowledge transferred from human demonstrations, in *Proceedings of 34th Chinese Control Conference*, Hangzhou, China, July 2015, pp. 3033–3038
9. S. Karakovskiy, J. Togelius, The mario ai benchmark and competitions. IEEE Trans Comput. Intell. AI Games **4**(1), 55–67 (2012)
10. S. Ontanón, G. Synnaeve, A. Uriarte, F. Richoux, D. Churchill, M. Preuss, A survey of real-time strategy game ai research and competition in starcraft. IEEE Trans. Comput. Intell. AI Games **5**(4), 293–311 (2013)
11. B. Takacs, J. Holaza, J. Stevek, M. Kvasnica, Export of explicit model predictive control to python, in *Proceedings of 20th International Conference on Process Control*, Štrbské Pleso, Slovakia, June 2015, pp. 78–83
12. A. Sahin, E. Atici, T. Kumbasar, Type-2 fuzzified flappy bird control system, in *Proceedings of IEEE International Conference on Fuzzy Systems* (2016), pp. 1578–1583
13. M. Ebner, J. Levine, S.M. Lucas, T. Schaul, T. Thompson, J. Togelius, Towards a video game description language (2013). https://doi.org/10.4230/dfu.vol6.12191.85
14. A. Sahin, T. Kumbasar, Landing on the moon with type-2 fuzzy logic, in *Proceedings of IEEE International Conference on Fuzzy Systems* (2017), pp. 1–6
15. J.M. Mendel, R.B. John, Type-2 fuzzy sets made simple. IEEE Trans. Fuzzy Syst. **10**(2), 117–127 (2002)
16. J.M. Mendel, H. Hagras, W.W Tan, W.W. Melek, H. Ying, *Introduction to Type-2 Fuzzy Logic Control* (John Wiley and IEEE Press, Hoboken, NJ, 2014)
17. H. Hagras, A hierarchical type-2 fuzzy logic control architecture for autonomous mobile robots. IEEE Trans. Fuzzy Syst. **12**(4), 524–539 (2004)
18. T. Kumbasar, H. Hagras, Big Bang–Big Crunch optimization based interval type-2 fuzzy pid cascade controller design strategy. Inform. Sci. **282**, 277–295 (2014)
19. A. Kumar, V. Kumar, Evolving an interval type-2 fuzzy PID controller for redundant robotic manipulator. Expert Syst. Appl. **73**, 161–177 (2017)

20. T. Kumbasar, H. Hagras, A self-tuning zSlices based general type-2 fuzzy PI controller. IEEE Trans. Fuzzy Syst. **23**(4), 991–1013 (2015)
21. M. Mehndiratta, E. Kayacan, T. Kumbasar, Design and experimental validation of single input type-2 fuzzy PID controllers as applied to 3 DOF helicopter testbed, in *Proceedings of IEEE International Conference on Fuzzy Systems* (2016), pp. 1584–1591
22. C. Lynch, H. Hagras, V. Callaghan, Embedded type-2 FLC for real-time speed control of marine and traction diesel engines, in *Proceedings of IEEE International Conference on Fuzzy Systems* (2005), pp. 347–352
23. E. Yesil, Interval type-2 fuzzy PID load frequency controller using Big Bang–Big Crunch optimization. Appl. Soft Comput. **15**, 100–112 (2014)
24. Q. Liang, J.M. Mendel, Interval type-2 fuzzy logic systems: theory and design. IEEE Trans. Fuzzy Syst. **8**(5), 535–550 (2000)
25. H. Wu, J.M. Mendel, Uncertainty bounds and their use in the design of interval type-2 fuzzy logic systems. IEEE Trans. Fuzzy Syst. **10**(5), 622–639 (2002)
26. T. Kumbasar, H. Hagras, Interval type-2 fuzzy PID controllers, in *Springer Handbook of Computational Intelligence* (Springer, Berlin, Heidelberg, 2015), pp. 285–294
27. T. Kumbasar, A simple design method for interval type-2 fuzzy PID controllers. Soft. Comput. **18**(7), 1293–1304 (2014)
28. T. Kumbasar, Robust stability analysis and systematic design of single input interval type-2 fuzzy logic controllers. IEEE Trans. Fuzzy Syst. **24**(3), 675–694 (2015). https://doi.org/10.1109/TFUZZ.2015.2471805
29. A. Taskin, T. Kumbasar, An open source Matlab/Simulink toolbox for interval type-2 fuzzy logic systems, in *Proceedings of IEEE Symposium Series on Computational Intelligence* (2015), pp. 1561–1568
30. A. Sakalli, T. Kumbasar, M.F. Dodurka, E. Yesil, The simplest Interval type-2 fuzzy PID controller: structural analysis, in *Proceedings of IEEE International Conference on Fuzzy Systems* (2014), pp. 626–633
31. A. Sakalli, T. Kumbasar, On the fundamental differences between the NT and the KM center of sets calculation methods on the IT2-FLC performance, in *Proceedings of IEEE International Conference on Fuzzy Systems* (2015), pp. 1–8
32. A. Sakalli, A. Bekea, T. Kumbasar, Gradient descent and extended kalman filter based self-tuning interval type-2 fuzzy PID controllers, in *Proceedings of IEEE International Conference on Fuzzy Systems* (2016), pp. 1592–1598
33. A. Isaksen, D. Gopstein, A. Nealen, Exploring game space using survival analysis, in *Foundations of Digital Games* (2015)
34. M. Zhang, *Flappy bird for MATLAB*, http://www.mathworks.com/matlabcentral/45795-flappy-bird-for-matlab. 25 Oct 2015
35. W.-C. So, C.K. Tse, Y.-S. Lee, Development of a fuzzy logic controller for dc–dc converters: design, computer simulation, and experimental evaluation. IEEE Trans. Power Electron. **11**(1), 24–32 (1996)
36. B. Edwards, Forty years of Lunar Lander, Technologizer (2009), http://www.technologizer.com/2009/07/19/lunar-lander/. Accessed 02 Jan 2017
37. S. Samothrakis, S.A. Roberts, D. Perez, S.M. Lucas, Rolling horizon methods for games with continuous states and actions, in *Proceedings of IEEE Conference on Computational Intelligence in Games* (2014), pp. 1–8
38. S.A. Roberts, S.M. Lucas, Measuring interestingness of continuous game problems, in *Proceedings of IEEE Conference on Computational Intelligence in Games* (2013) pp. 1–8
39. H. Corte, Moon Lander matlab game (2012), https://uk.mathworks.com/matlabcentral/fileexchange/38927. Accessed 02 Jan 2017

A Type-2 Fuzzy Model to Prioritize Suppliers Based on Trust Criteria in Intelligent Agent-Based Systems

Mohammad Hossein Fazel Zarandi, Zohre Moattar Husseini
and Seyed Mohammad Moattar Husseini

Abstract In the last two decades the intelligent agents have improved the lifestyle of human beings from different aspects of view such as life activities and services. Considering the importance of the safety and security role in the e-procurement, there have been many systems developed including trust engine. In particular, some of the first systems were modeled though trust evaluation concepts as crisp values, but now a days to adjust the systems with real world cases, the uncertainty and impreciseness parameters must be considered with the use of fuzzy sets theory. In this paper to minimize the number of exceptions related to suppliers, Trust Management Agent (TMA) is considered to prioritize candidate suppliers based on trust criteria. Due to lots of uncertainties, type-2 fuzzy sets prove to be a most suitable methodology to deal with the trust evaluation process efficiently. In this regard, a new evaluation process based on hierarchical Linguistic Weighted Averaging (LWA) sets is proposed. The solution method was then illustrated through a simple example which clarifies the suitability as well as the simplicity of the proposed method for the category of the defined problem.

Keywords Interval type-2 fuzzy · Intelligent agent-based systems
Trust evaluation · E-procurement · Prioritize suppliers

1 Introduction

The term e-Procurement refers to the use of electronic communications to deal with business process between sellers and buyers, through linking and integrating inter-organization business processes and systems with the use of Internet-based

M. H. Fazel Zarandi (✉) · Z. Moattar Husseini · S. M. Moattar Husseini
Department of Industrial Engineering and Management System,
Amirkabir University of Technology, Tehran, Iran
e-mail: zarandi@aut.ac.ir

Z. Moattar Husseini
e-mail: z.moattar@aut.ac.ir

© Springer International Publishing AG 2018   129
R. John et al. (eds.), *Type-2 Fuzzy Logic and Systems*,
Studies in Fuzziness and Soft Computing 362,
https://doi.org/10.1007/978-3-319-72892-6_7

protocols [2]. E-Procurement has been providing more efficient trading methods as well as new trading opportunities in the supply networks.

In the last two decades the intelligent agents have improved the lifestyle of human beings from different aspects including life activities and services. There has been growing interest in the design of a distributed, intelligent society of agents in e-commerce applications in the recent years [10, 13, 21, 30].

It is crucial, in an agent based e-procurement system to protect both buyers and sellers from any possible unsatisfied condition, which is commonly due to some uncertain and vague characters. In this regard, there have been a number of systems developed using a kind of trust engine to help establish trust orientation between the firms [33]. Such trust, could positively affect firms' behaviors and performances and meantime reduce their interrelation risks.

The establishment of the trust commonly requires one party to assess the other on its past behaviors, acts and promises based on some appropriate trust criteria [7, 22, 32]. For this assessment, commonly, not crisp but vague and uncertain data are available. Further, in the agent based systems, the assessment heavily relies on the collective opinions from the agents in the community. Whilst, some early agent based systems modeled the trust evaluation process merely using crisp values, but to adjust the systems with real world cases, the uncertainty and vagueness parameters must be considered in the modelling through utilizing fuzzy based theories.

While type-1 fuzzy sets are capable to handle several kinds of uncertainties [16] these are not able to directly model uncertainties related to some particular sources, such as: uncertainty in the meanings of the words and uncertainty associated with the consequences (e.g. when the knowledge extracted from group of experts who do not all agree). Type-2 fuzzy sets are more appropriate for these situations. Type-2 fuzzy sets utilizes higher degree of freedom by a fuzzy membership functions to handle uncertainties in real world situations.

Considering uncertainty characteristics of the inter-organizational trust evaluation in an agent based e-procurement system (as detailed in the next section) and based on the capabilities outlined for type-2 fuzzy sets (in Sect. 4) this paper propose a new evaluation process based on Linguistic Weighted Averaging (LWA) sets using Interval Type-2 Fuzzy set (IT2-FSs).

The following section (Sect. 2) provides a brief review of the literature and the main subjects concerned in this paper. Section 3 describes the defined problem. The background of the solution approach is presented in Sect. 4 and the solution approach is detailed in Sect. 5. Then an illustrative example is presented in Sect. 6. Finally, conclusions are provided in Sect. 7.

2 Literature Review

This section provides brief reviews for the three main aspects concerned in this work:

- Intelligent agents in e-procurement
- Inter-organizational trust in the e-procurement intelligent agents
- The capabilities of T2-FSs in dealing with high levels of uncertainties.

2.1 Intelligent Agents in E-Procurement

E-Procurement is considered to be a strategic tool for improving the competitiveness of organizations and generating scale economies for both sellers and buyers. In this context, one critical issue is to tackle problems existing in ensuring a trustworthy environment in which business interrelationship risks can be minimized [2].

Intelligent agents reveal the capability to operate on behalf of buyers to look for requested products concerning the process of procurement [9]. Raghavan and Prabhu [19] developed a software for agent-based framework considering a typical e-procurement process by classifying the procurement process into three classes: e-negotiations, e-settlement, and reverse auctions [19]. Cheung et al. [6] proposed an agent-oriented knowledge-based system for strategic e-procurement using real time information to produce dynamic business rules [6]. Lee and his collogues [12] proposed an agent based e-procurement system, in which the intelligent agents are responsible for searching and negotiating the potential suppliers and evaluating the performance of suppliers based on the selection criteria [12, 26].

Sun and his collogues [26] proposed an agent and Web service based architecture for considering exception handling in e-procurement. In this architecture, different tasks in the e procurement process are assigned to different agents, such as searching, negotiating, supplier selection, contracting, monitoring, and exception handling [26].

Despite the existing developments on applying intelligent agents for e-procurement, the challenge remains on how to tackle the existing problem as the legal framework that can ensure a trustworthy environment (as mentioned in this section). For this reason Inter organizational trust in intelligent agents of e-procurement is considered in the next sub-section.

2.2 Inter-organizational Trust in the E-Procurement Intelligent Agents

Inter-organizational trust helps establish a kind of inter-firm relationship which ensures each side holding a collective trust orientation towards the other [33]. This positively affects firms' behaviors and performances and meantime reduces interrelation risks. Inter-organizational trust is conceptualized as a multi-dimensional construct, for which a list of 22 widely referred dimensions is introduced in [25].

The same paper also summarized the most commonly used dimensions as: credibility, benevolence, goodwill, predictability, reciprocity, openness and confidence.

Trust has been recognized as a key issue in multi-agent and e-commerce systems, being at the core of the interactions between agents operating in uncertain business environments [11, 20]. Bases of the trustworthiness knowledge is one main concern, for which three types are commonly agreed in the literature: individual experience, inference from other agents in the community, and a hybrid of the two [3]. Evidences used for trust evaluation, based on their source types, can also be categorized as priori evidences and experienced evidences [11, 18, 35]. Priori evidences are those mainly provided by protocols, policies, or mechanisms; while experienced evidences are obtained by the agents during their interactions. The literature reveals considerable research interest in trust decision with regard to the community based experiences. With regard to the characteristics of trust evaluation in the e-procurement environment, we refer to [1] which states the need to deal with high levels of uncertainties, vagueness and ambiguities which are commonly due to: (1) The absence of an authority to prescribe the rules for inter-organizational interaction as buyer supplier relationships, (2) Trading transactions might occur among unknown parties, which requires a collection of indirect trust experience from referee agents in the community, and (3) The use of trust experiences which are based on the feedback from buyers.

According to some studies [20, 27], for an agent to evaluate other agents' trustworthiness some models traditionally use a bi-stable value (good or bad), while this cannot generally support realistic situations. Instead, some other researches (e.g. [18, 23, 24]) attribute some fuzziness to the notion of performance and then evaluate the trustworthiness using fuzzy reasoning techniques. These authors concluded that the fuzzy reasoning is especially attractive for the trust evaluation purpose.

The above review leads to the recognition of the characteristics of trust evaluation in the e-procurement context the characteristic represent the high levels of uncertainties, vagueness and ambiguities (as detailed [1]).

2.3   A Short Review of T2-FSs Capabilities

In 1975 Zadeh introduced type-2 fuzzy sets to minimize the effect of uncertainties concerning ambiguity, vagueness and randomness [31]. While type-1 fuzzy sets are capable to handle several kinds of uncertainties, according to Mendel and his colleagues in [16] these are not able to directly model uncertainties related to the following sources: (1) Uncertainty in the meanings of the terms (for instance used in the rules), (2) Uncertainty associated with the consequences (for instance when the knowledge extracted from group Of experts who do not all agree), (3) Uncertainty in the measurements that activate type-1 fuzzy set and (4) Uncertainty in the data used to tune parameters of a type-1 fuzzy sets. These types of uncertainties all translate into uncertainties about fuzzy sets membership functions, while in the type-1 fuzzy sets membership functions are totally crisp. In this paper, it is concluded that type-2 fuzzy sets are able to model such types of uncertainties because of fuzzy membership functions. According to Castillo and his colleagues [5] five types of uncertainties emerge from imprecise knowledge natural state, which are: uncertainties related to measurement, process, model, estimate and implementation.

As discussed in [34] a type-2 fuzzy set, which is characterized by a fuzzy membership function, is capable to provide us with more degrees of freedom to represent the vagueness and the uncertainty and of the real world.

According to Mendel and his colleagues [16], there has been several application areas for fuzzy logic systems and type-2 fuzzy sets (for instance decision making, extracting knowledge from questionnaire surveys, function approximation, learning linguistic membership grades, preprocessing radiographic images and transport scheduling).

3 Problem Description

As discussed in the previous section, it is crucial in an agent based e-procurement system to protect the buyer from any possible unsatisfied condition which is commonly due the existence of uncertain and vague characters. In this regard, many research attempts have been reported in the literature including some developments which use an exception management agent to handle such undesired situations. Some research works with the inter-organizational trust orientation have been also presented in the literature. It is notable that presence of the trust in the buyer-seller relationship not only reduces uncertainty and vagueness characteristics, but significantly reduces the complexity of the inter-firm relations which could in turn enhance their trading process. These published works, however, mostly utilize T1-FSs in their solution approach, therefore, they provide limited capabilities in handling mentioned uncertainty and vagueness characteristics. As reviewed in the literature, T2-FSs utilize higher degree of freedom by a fuzzy membership function to handle uncertainties and vagueness in real world situations.

Based on the recognition of the above remarking points, the current paper aims to establish inter-organizational trust in the agent based e-procurement systems, through proposing an effecting supplier evaluation and ranking method. Considering previously mentioned characteristics inherent in the inter-organizational trust evaluation process such as; uncertainty and vagueness in the data, use of collective opinions from experts or other agent in the community, as well as using both direct and indirect sources of evidences; this paper utilizes type-2 fuzzy sets as the solution approach for the defined problem.

The paper considers an agent based e-procurement system consisting of specific agents for the required functions also including a Trust Management Agent (ATM) which is responsible to establish inter-organizational trust in the buyer-supplier relationship. One major function in this respect is to evaluate some pre-qualified candidate suppliers in order to rank them on some particular trust criteria. In this regard the paper aims to propose a solution method to determine various decision to lead to a short list of the most appropriate suppliers ranked based on their trustworthiness characteristics. It is notable that this approach utilizes linguistic weighted averaging based on interval type-2 fuzzy sets in the evaluation process.

The proposed solution is further detailed in the following sections.

4 Basic Concepts of Type-2 Fuzzy Sets

In 1975 Zadeh introduced type-2 fuzzy sets to minimize the effect of uncertainties concerning ambiguity, vagueness and randomness [31]. Comparing to an ordinary (type-1) fuzzy set which has a grade of crisp membership function, a type-2 fuzzy set has grades of fuzzy membership functions [14]. This section is organized to review theoretical definitions related to the proposed fuzzy type-2 based solution method, including: Interval Type-2 Fuzzy Sets (IT2-FSs) and Linguistic Weighted Averaging (LWA).

4.1 Interval Type-2 Fuzzy Sets

Definition 1 [16] A type-2 fuzzy set $\tilde{A}$ is characterized by a type-2 membership function $\mu_{\tilde{A}}(x, u)$, where $x \in X$, $u \in J_x \subseteq [0, 1]$ and $\mu_{\tilde{A}}(x, u) \subseteq [0, 1]$, i.e.,

$$\tilde{A} = \{((x, u),\ \mu_{\tilde{A}}(x, u)) | \forall x \in X,\ \forall u \in J_x \subseteq [0, 1]\} \tag{1}$$

Also $\tilde{A}$ can be presented by Eq. (2),

$$\tilde{A} = \int_{x \in X} \int_{u \in J_x} \mu_{\tilde{A}}(x, u) / (x, u)\ J_x \subseteq [0, 1] \tag{2}$$

where $\int$ denotes union overall admissible x and u. For discrete universe of discourse x and u, $\int$ is replaced by $\sum$.

Definition 2 [17] If all $\mu_{\tilde{A}}(x, u) = 1$ then $\tilde{A}$ is an interval type-2 fuzzy sets which can be expressed as a special case of general type-2 fuzzy sets, Eq. (3):

$$\tilde{A} = \int_{x \in X} \int_{u \in J_x} 1 / (x, u) \qquad J_x \subseteq [0, 1]. \tag{3}$$

Note that x is the primary variable, $J_x \subseteq [0, 1]$ is the primary MF of x, also u is the secondary variable, and $\int_{u \in J_x} 1/u$ is the secondary MF at x.

Definition 3 [17] A bounded region with respect to the uncertainty in the primary memberships of an IT2-FS, is called the Footprint of Uncertainty (FOU), which is the union of all primary membership functions.

$$FOU(\tilde{A}) = \bigcup_{x \in X} J_x. \tag{4}$$

So, FOU demonstrates the vertical-slice-representation to indicate the interval type-2 fuzzy sets.

Definition 4 [28] The FOU is bounded by an Upper Membership Function (UMF) $\bar{A}(x) \equiv \bar{A}$ and a Lower Membership Function (LMF) $\underline{A}(x) \equiv \underline{A}$, which are T1-FSs; So, the membership function of each element of an IT2-FS is an interval $[\underline{A}(x), \bar{A}(x)]$.

4.2 Linguistic Weighted Averaging

The linguistic weighted average, concerning IT2-FSs as inputs, is introduced by Wu and Mendel in 2007 and 2008 in [28, 29] which is an extension of the Fuzzy Weighted Average (FWA) [8] for type-1 FSs inputs. The LWA is defined:

$$\tilde{Y}_{LWA} = \frac{\sum_{i=1}^{n} \tilde{X}_i \tilde{W}_i}{\sum_{i=1}^{n} \tilde{W}_i} \tag{5}$$

where $\tilde{X}_i$ and the corresponded weight $\tilde{W}_i$ are linguistic terms. Considering that $\tilde{X}_i$ and $\tilde{W}_i$ which are modeled by IT2-FSs, the $\tilde{Y}_{LWA}$ is also IT2-FSs (Eq. (6)),

$$\tilde{Y}_{LWA} = 1/FOU(\tilde{Y}_{LWA}) = 1/[\underline{Y}_{LWA}, \bar{Y}_{LWA}] \tag{6}$$

where $\underline{Y}_{LWA}$ and $\bar{Y}_{LWA}$ are LMFs and UMFs of $\tilde{Y}_{LWA}$, respectively [28]. Considering the use of $\tilde{X}_i$ and $\tilde{W}_i$ in computing $\tilde{Y}_{LWA}$, with regard to vertical-slice-representation, Eqs. (7) and (8) are defined as below [28]:

$$\tilde{X}_i = 1/FOU(\tilde{X}_i) = 1/[\underline{X}_i, \bar{X}_i] \tag{7}$$

$$\tilde{W}_i = 1/FOU(\tilde{W}_i) = 1/[\underline{W}_i, \bar{W}_i] \tag{8}$$

where $\underline{X}_i$ and $\bar{X}_i$ ($\underline{W}_i$ and $\bar{W}_i$) are LMFs and UMFs of $\tilde{X}_i(\tilde{W}_i)$, respectively.

$\bar{Y}_{LWA}$ and $\underline{Y}_{LWA}$ will be computed using the α-cut, in which the range of the MF is discretized into m points as $\alpha_1, \alpha_2, \ldots, \alpha_m$. The α-cut on $\tilde{X}_i$ and $\tilde{W}_i$ are applied to compute the corresponding $\tilde{Y}_{LWA}$ (Figs. 1, 2 and 3) [28].

Fig. 1 $\tilde{X}_i$ and an α-cut. the dashed curve is an embedded T1 FS of $\tilde{X}_i$ [28]

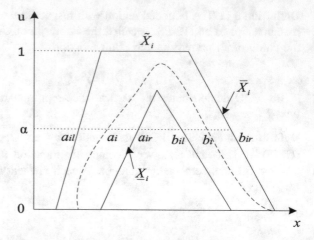

Fig. 2 $\tilde{W}_i$ and an α-cut. The dashed curve is an embedded T1 FS of $\tilde{W}_i$ [28]

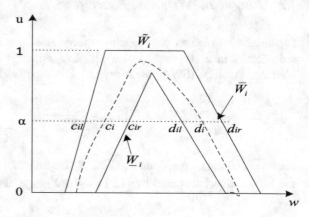

Fig. 3 $\tilde{Y}_i$ and an α-cut [28]

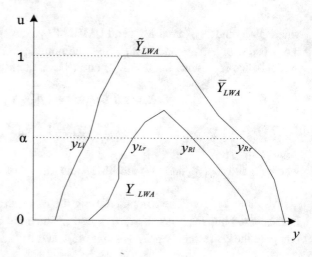

Noted that all UMFs are T1 FSs normal, so the height of the UMFs of $\bar{Y}_{LWA}$ is one ($h_{\bar{Y}_{LWA}} = 1$).

And the height of $\underline{Y}_{LWA}$ which is the lower bound of MFs of FOU ($\tilde{Y}_{LWA}$) is calculated by h_{min} which is defined as the smallest height of all FWAs resulted from T1 FSs of the height of $\underline{X}_i$ as $h_{\underline{X}_i}$ and $\underline{W}_i$ as $h_{\underline{W}_i}$ in Eq. (9) [14].

$$h_{min} = \min\{\min_{\forall i} h_{\underline{X}_i}, \min_{\forall i} h_{\underline{W}_i}\} \tag{9}$$

Let the interval of $[a_i(\alpha), b_i(\alpha)]$ be an α-cut on $\tilde{X}_i$, and the interval $[c_i(\alpha), d_i(\alpha)]$ be an α-cut on $\tilde{W}_i$. As shown in Fig. (1), if the α-cut on $\underline{X}_i$ exists, then the interval $[a_{il}(\alpha), b_{ir}(\alpha)]$ is divided into three subintervals: $[a_{il}(\alpha), a_{ir}(\alpha)]$, $[a_{ir}(\alpha), b_{il}(\alpha)]$, and $[b_{il}(\alpha), b_{ir}(\alpha)]$.

However, if the α on $\underline{X}_i$ is larger than $h_{\underline{X}_i}$ (α-cut dose not exist), then both the value of $a_i(\alpha)$ and $b_i(\alpha)$ can be assumed in the entire interval $[a_{il}(\alpha), b_{ir}(\alpha)]$:

$$a_i(\alpha) = \begin{cases} [a_{il}(\alpha), a_{ir}(\alpha)] & \alpha \in [0, h_{\underline{X}_i}] \\ [a_{il}(\alpha), b_{ir}(\alpha)] & \alpha \in [h_{\underline{X}_i}, 1] \end{cases} \tag{10}$$

$$b_i(\alpha) = \begin{cases} [b_{il}(\alpha), b_{ir}(\alpha)] & \alpha \in [0, h_{\underline{X}_i}] \\ [a_{il}(\alpha), b_{ir}(\alpha)] & \alpha \in [h_{\underline{X}_i}, 1] \end{cases} \tag{11}$$

Similarly the value of $c_i(\alpha)$ and $d_i(\alpha)$ can be assumed based on Fig. 2,

$$c_i(\alpha) = \begin{cases} [c_{il}(\alpha), c_{ir}(\alpha)] & \alpha \in [0, h_{\underline{W}_i}] \\ [c_{il}(\alpha), d_{ir}(\alpha)] & \alpha \in [h_{\underline{W}_i}, 1] \end{cases} \tag{12}$$

$$d_i(\alpha) = \begin{cases} [d_{il}(\alpha), d_{ir}(\alpha)] & \alpha \in [0, h_{\underline{W}_i}] \\ [c_{il}(\alpha), d_{ir}(\alpha)] & \alpha \subset [h_{\underline{W}_i}, 1] \end{cases} \tag{13}$$

In Eqs. (10)–(13), the l and r *are the left and* right indices respectively. Also the value of $a_{ir}(\alpha), b_{il}(\alpha), c_{il}(\alpha)$ and $d_{il}(\alpha)$ can be defined as Eqs. (14)–(17):

$$a_{ir}(\alpha) \triangleq \begin{cases} a_{ir}(\alpha), & \alpha \leq h_{\underline{X}_i} \\ b_{ir}(\alpha), & \alpha > h_{\underline{X}_i} \end{cases} \tag{14}$$

$$b_{il}(\alpha) \triangleq \begin{cases} b_{il}(\alpha), & \alpha \leq h_{\underline{X}_i} \\ a_{il}(\alpha), & \alpha > h_{\underline{X}_i} \end{cases} \tag{15}$$

$$c_{ir}(\alpha) \triangleq \begin{cases} c_{ir}(\alpha), & \alpha \leq h_{\underline{W}_i} \\ d_{ir}(\alpha), & \alpha > h_{\underline{W}_i} \end{cases} \tag{16}$$

$$d_{il}(\alpha) \triangleq \begin{cases} d_{il}(\alpha), & \alpha \leq h_{\underline{W}_i} \\ c_{il}(\alpha), & \alpha > h_{\underline{W}_i} \end{cases} \tag{17}$$

In the LWA, the value of $a_i(\alpha)$, $b_i(\alpha)$, $c_i(\alpha)$, and $d_i(\alpha)$ can be assumed continuously in their corresponding α-cut intervals. So numerous different combinations of those values can be produced to form $y_L(\alpha)$ and $y_R(\alpha)$. By considering all $y_L(\alpha)$ and $y_R(\alpha)$, continuous intervals $[y_{Ll}(\alpha), y_{Lr}(\alpha)]$ and $[y_{Rl}(\alpha), y_{Rr}(\alpha)]$ are obtained, where $y_{Lr}(\alpha)$, $y_{Rl}(\alpha)$, $y_{Ll}(\alpha)$, and $y_{Rr}(\alpha)$ are illustrated in (Fig. 3):

$$\underline{Y}_{LWA}(\alpha) = [y_{Lr}(\alpha), y_{Rl}(\alpha)], \quad \alpha \in [0, h_{\min}] \tag{18}$$

$$\bar{Y}_{LWA}(\alpha) = [y_{Ll}(\alpha), y_{Rr}(\alpha)], \quad \alpha \in [0, 1] \tag{19}$$

Considering the fix values of $a_i(\alpha)$, $b_i(\alpha)$, $c_i(\alpha)$ and $d_i(\alpha)$, the values of $y_{Ll}(\alpha)$, $y_{Lr}(\alpha)$, $y_{Rl}(\alpha)$ and $y_{Lr}(\alpha)$ are defined as below [14, 28, 29]:

$$y_{Ll}(\alpha) = \frac{\sum_{i=1}^{L_l^*} a_{il}(\alpha)d_{ir}(\alpha) + \sum_{i=L_l^*+1}^{n} a_{il}(\alpha)c_{il}(\alpha)}{\sum_{i=1}^{L_l^*} d_{ir}(\alpha) + \sum_{i=L_l^*+1}^{n} c_{il}(\alpha)} \quad \alpha \in [0, 1] \tag{20}$$

$$y_{Lr}(\alpha) = \frac{\sum_{i=1}^{L_r^*} a_{ir}(\alpha)d_{il}(\alpha) + \sum_{i=L_r^*+1}^{n} a_{ir}(\alpha)c_{ir}(\alpha)}{\sum_{i=1}^{L_r^*} d_{il}(\alpha) + \sum_{i=L_r^*+1}^{n} c_{ir}(\alpha)} \quad \alpha \in [0, h_{\min}] \tag{21}$$

$$y_{Rl}(\alpha) = \frac{\sum_{i=1}^{R_l^*} b_{il}(\alpha)c_{ir}(\alpha) + \sum_{i=R_l^*+1}^{n} b_{il}(\alpha)d_{il}(\alpha)}{\sum_{i=1}^{R_l^*} c_{ir}(\alpha) + \sum_{i=R_l^*+1}^{n} d_{il}(\alpha)} \quad \alpha \in [0, h_{\min}] \tag{22}$$

$$y_{Rr}(\alpha) = \frac{\sum_{i=1}^{R_r^*} b_{ir}(\alpha)c_{il}(\alpha) + \sum_{i=R_r^*+1}^{n} b_{ir}(\alpha)d_{ir}(\alpha)}{\sum_{i=1}^{R_r^*} c_{il}(\alpha) + \sum_{i=R_r^*+1}^{n} d_{ir}(\alpha)} \quad \alpha \in [0, 1] \tag{23}$$

In these Equations, L_l^*, L_r^*, R_l^* and R_r^* are defined as switch points which are computed by KM or EKM algorithms discussed in [15]. $y_{Ll}(\alpha)$ and $y_{Rr}(\alpha)$ as shown in Figs. 1, 2 and 3 and Eqs. (20) and (23) only depend on the UMFs of $\tilde{X}_i$ and $\tilde{W}_i$, which are computed from the corresponding α-cuts (Expressive Eq. (24)).

$$\bar{Y}_{LWA} = \frac{\sum_{i=1}^{n} \bar{X}_i \bar{W}_i}{\sum_{i=1}^{n} \bar{W}_i} \tag{24}$$

Because all $\bar{X}_i$ and $\bar{W}_i$ are normal T1-FSs, $\bar{Y}_{LWA}$ is also normal.

Similarly, observe from Eqs. (21) and (22) and the mentioned Figures, the $y_{Lr}(\alpha)$ and $y_{Rl}(\alpha)$ only depend on the LMFs of $\tilde{X}_i$ and $\tilde{W}_i$, which are computed from the corresponding α-cuts (Expressive Eq. (25)).

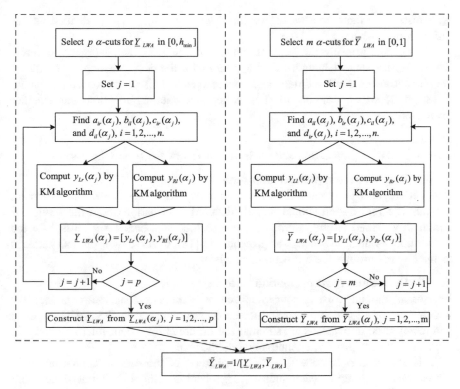

Fig. 4 The pseudo-code of computing LWA [29]

$$\underline{Y}_{LWA} = \frac{\sum_{i=1}^{n} \underline{X}_i \underline{W}_i}{\sum_{i=1}^{n} \underline{W}_i} \qquad (25)$$

Noted that, unlike $\bar{Y}_{LWA}$, the height of $\underline{Y}_{LWA}$ is h_{min}, which is defined by Eq. (9), as the minimum height of all $\underline{X}_i$ and $\underline{W}_i$.

A pseudo-code for computing $\bar{Y}_{LWA}$ and $\underline{Y}_{LWA}$ is given in Fig. 4.

5 Solution Approach

This section describes the proposed method of prioritizing suppliers based on trust criteria in the intelligent agent environment considering type-2 fuzzy sets.

Let $S_i = \{s_1, s_2, \ldots, s_z\}$ be a set of prequalified (candidate) suppliers. A set of agents $A_f = \{a_1, \ldots, a_m\}$ is considered to query from, also concerning their frequency of interaction with suppliers, linguistic weights $WA_f = \{wa_1, \ldots, wa_m\}$ are defined, where $\sum_{f=1}^{m} WA_f = 1$. Also $C_h = \{c_1, c_2, \ldots, c_n\}$ is a set of trust criteria with respect to their importance weights $W_h = \{w_1, w_2, \ldots, w_n\}$, defined as linguistic terms, where $\sum_{h=1}^{n} W_h = 1$.

The supplier prioritization process consists of 6 Steps, which is organized in two Stages. While, stage 1 includes three functions for the determination of: candidate suppliers, trust evaluation criteria and referee agents. This paper concentrates more on stage 2, which consists of data collection, data aggregation and supplier prioritization.

Stage 1

This is a kind of preparation stage which consists of the following 3 steps.

Step 1: Determine prequalified suppliers

Suppliers register their services in the service register center of the e-procurement system. The registration should include detailed information about the service or commodity. Based on the information, prequalified suppliers are identified for the further trust evaluation process, as presented in [18], to serve their product catalogs to the buyers through the e-procurement system

Step 2: Determine trust evaluation criteria

Inter-organizational trust is conceptualized as a multi-dimensional construct, for which a list of 22 widely referred dimensions are reviewed in [25] while the most commonly used dimensions are summarized as: credibility, benevolence, goodwill, predictability, reciprocity, openness and confidence.

This Step is concerned with the determination of those more appropriate trust dimensions considering each industry/organization situation. Furthermore, with regard to the firm's strategic aspects, an appropriate weight has also to be considered for each selected criterion.

Step 3: Determine referee agents

Multi-agent systems work autonomously and collaboratively by mean of the Internet. Each agent are focused on its own particular tasks, meanwhile cooperatively provide a specific operation or service to other agents [26]. To retrieve information from agents, a set of agents who has historical data about direct and indirect trust experience with the target suppliers is identified.

Stage 2

This stage consists of the following three steps to prioritize suppliers based on the trust criteria concerning interval type-2 fuzzy sets (both data judgment and weights).

Step 1: Data collection

When the supplier prioritization is required, the TMA queries about suppliers, concerning trust criteria, from each of the referee agent. These agents are asked to evaluate the suppliers by completing an electronic form which is provided by TMA as shown in Fig. 5.

The evaluation of the agents are based on the defined trust criteria as mentioned in Stage 1–Step 2. There are different trust assessment levels which are ranged from

Assessment	Weakly Trustable		Moderately Trustable		Strongly Trustable		Extremely Trustable	
Trust criteria 1 (C_1)	( )		( )		( )		( )	
Trust criteria 2 (C_2)	( )		( )		( )		( )	
Trust criteria n (C_n)	( )		( )		( )		( )	
Term of interaction	Never	Almost Never	Seldom	Unspecified	Often	Almost always	Always	
Frequency of interaction	( )	( )	( )	( )	( )	( )	( )	

Fig. 5 The trust evaluation form of ith supplier for the agent f

the worst to the best based on linguistic terms as: Weakly Trustable (WT), Moderately Trustable (MT), Strongly Trustable (ST) and Extremely Trustable (ET). Also weights of the frequency of interaction associated with agents and suppliers are concerned using linguistic terms as: Never (N), Almost Never (AN), Seldom (S), Unspecified (U), Often (O), Almost Always (AA) and Always (A). in this paper, it is assumed that each category of terms, as mentioned, has been explained as IT2 FSs $\tilde{X}$, using upper and lower fuzzy membership functions (Tables 1 and 2).

Step 2: Data aggregation
In this paper two phases of aggregation are defined as below:

- **Step 2-1**: The aggregation process with respect to agents' judgments.
  In this phase linguistic judgments of agents are aggregated for each criterion and supplier. Considering that the frequency of interaction between agents and suppliers can improve the accuracy of the collected data, the LWA operator is applied to aggregate the all agents' judgment for the hth criterion of the ith supplier with regard to the linguistic data related to frequency of interaction. The frequency of interaction of agents can be considered based on the filled form as shown in Fig. 5. So in this case, the opinion of the agent who always interact with specific supplier is taken into account by considering higher linguistic weights.
- **Step 2-2**: The aggregation process with respect to defined trust criteria.

Table 1 Fuzzy membership functions of agents' judgments based on linguistic terms

Linguistic variables	Fuzzy type 1	Fuzzy type 2	
	Membership function (MFs)	Upper membership function (UMF)	Lower membership function (LMF)
Weakly Trustable (WT)	(0, 2, 4)	(0.00, 2, 4.2)	(0.20, 2, 3.80)
Moderately Trustable (MT)	(2, 4, 6)	(1.80, 4, 6.2)	(2.20, 4, 5.80)
Strongly Trustable (ST)	(4, 6, 8)	(3.80, 6, 8.2)	(4.20, 6, 7.80)
Extremely Trustable (ET)	(6, 8, 10)	(5.80, 8, 10)	(6.20, 8, 9.80)

Table 2 Fuzzy membership functions assigned to agents' and criteria's weight based on linguistic terms

Linguistic variables of agents	Linguistic variables of criteria	Fuzzy type 1	Fuzzy type 2	
		Membership function(MFs)	Upper membership function (UMF)	Lower membership function (LMF)
Never (N)	Very Low (VL)	(0.1, 0.5, 1)	(0.06, 0.5, 1.05)	(0.14, 0.5, 0.95)
Almost Never (AN)	Low (L)	(0.5, 1, 3)	(0.45, 1, 3.2)	(0.55, 1, 2.80)
Seldom (S)	Medium Low (ML)	(1, 3, 5)	(0.80, 3, 5.2)	(1.20, 3, 4.80)
Unspecific (U)	Moderate (M)	(3, 5, 7)	(2.80, 5, 7.2)	(3.20, 5, 6.80)
Often (O)	Medium High (MH)	(5, 7, 8)	(4.80, 7, 8.1)	(5.20, 7, 7.9)
Almost always (AA)	High (H)	(7.5, 8, 9.5)	(7.45, 8, 9.65)	(7.55, 8, 9.35)
Always (A)	Very High (VH)	(9, 9.5, 10)	(8.95, 9.5, 10)	(9.05, 9.5, 9.95)

In this phase the aggregated data of previous Phase are used to aggregate them based on the trust criteria for each supplier. Considering the importance of each criterion, the TMA assigns an importance weight to each criterion based on the different assessment level that range from the lowest to the highest as defined by linguistic terms: Very Low (VL), Low (L), Medium Low (ML), Moderate (M), Medium High (MH), High (H) and Very High (VH).

In this paper assumed that each category of terms, as mentioned, has been explained as IT2 FSs $\tilde{X}$ using upper and lower fuzzy membership functions (Table 2). LWA operator is applied to aggregate the data of all criteria for each supplier based on the importance weight of the criterion.

Step 3—Suppliers prioritization

Different approaches to prioritizing/ranking interval type-2 fuzzy sets exist. In this paper, a method proposed by Asan and his colleagues in [4] is applied to rank suppliers based on α-cuts in the form of IT2-FSs in which both UMFs and LMFs are normal T1-FSs.

Let $Y_i^M(\alpha)$ in Eq. (26), denote the total mean of the end points crossing α-cuts on both LMFs and UMFs of $\tilde{Y}_i$,

$$Y_i^M(\alpha) = \frac{y_{Ll}(\alpha) + y_{Lr}(\alpha) + y_{Rl}(\alpha) + y_{Rr}(\alpha)}{4} \tag{26}$$

Also $|Y_i(\alpha)|$ in Eq. (27), considered as a weighting factor using the length of the α-cuts of the embedded average T1 FN.

$$|Y_i(\alpha)| = \frac{y_{Rl}(\alpha) + y_{Rr}(\alpha)}{2} - \frac{y_{Ll}(\alpha) + y_{Lr}(\alpha)}{2} \tag{27}$$

Then the ranking value rs_i of the supplier with respect to IT2-FSs $(\tilde{Y}_i)$ is calculated by Eq. (28) as proposed in [4]:

$$
\begin{aligned}
rs_i &= \frac{\int_0^1 Y_i^M(\alpha) |Y_i(\alpha)| d\alpha}{\int_0^1 |Y_i(\alpha)| d\alpha} \\
&= \frac{\int_0^1 \left(\frac{y_{Ll}(\alpha) + y_{Lr}(\alpha) + y_{Rl}(\alpha) + y_{Rr}(\alpha)}{4} \right) \left(\frac{y_{Rl}(\alpha) + y_{Rr}(\alpha)}{2} - \frac{y_{Ll}(\alpha) + y_{Lr}(\alpha)}{2} \right) d\alpha}{\int_0^1 \left(\frac{y_{Rl}(\alpha) + y_{Rr}(\alpha)}{2} - \frac{y_{Ll}(\alpha) + y_{Lr}(\alpha)}{2} \right) d\alpha}
\end{aligned}
\tag{28}
$$

In this case higher ranking value (rs_i) indicates more suitable supplier based on trust criteria compared to others.

6 Numerical Example

This section prioritizes the suppliers concerning interval type-2 fuzzy sets, using a simplified example. Based on the description of Stage1 in the previous section, a set of prequalified suppliers $S_i = \{s_1, s_2, \ldots, s_5\}$ is candidated by TMA as well as three referee agents $\{a_1, a_2, a_3\}$. Weights considered for the referee agents, with respect to their frequency of interaction toward target supplier, are $\{\widetilde{wa}_1, \widetilde{wa}_2, \widetilde{wa}_3\}$ as shown in Table 4. Four trust criteria $\{c_1, c_2, c_3, c_4\}$ for example: credibility, confidence, benevolence and predictability (as discussed in Sect. 1) are considered

Table 3 All agents' judgments for trustworthiness of the suppliers based on linguistic terms

Criterion	Importance weight of the criterion	Agents	Suppliers (Sup)				
			Sup 1	Sup 2	Sup 3	Sup 4	Sup 5
Credibility (Cr)	H	Agent 1	MT	MT	ET	MT	MT
		Agent 2	WT	WT	ET	MT	WT
		Agent 3	WT	ST	MT	MT	MT
Confidence (Con)	MH	Agent 1	WT	ST	ET	MT	WT
		Agent 2	MT	ST	ST	ST	WT
		Agent 3	MT	WT	ET	ST	MT
Benevolence (B)	M	Agent 1	WT	MT	ET	MT	MT
		Agent 2	WT	MT	ST	WT	MT
		Agent 3	MT	ST	ET	MT	WT
Predictability (P)	ML	Agent 1	WT	MT	ET	MT	WT
		Agent 2	WT	MT	ET	MT	MT
		Agent 3	WT	MT	MT	WT	MT

Table 4 Linguistic weights of agents based on frequency of interaction

Criterion	Agents		
	(Agent 1)	(Agent 2)	(Agent 3)
Supplier 1	U	O	O
Supplier 2	O	A	U
Supplier 3	O	O	AA
Supplier 4	U	U	U
Supplier 5	S	U	U

to evaluate the candidate suppliers. Importance weights associated with for the defined criterion 1 to 4 are $\{\tilde{w}_1, \tilde{w}_2, \tilde{w}_3, \tilde{w}_4\}$ respectively.

In this method, considering Step 1 of the Stage 2, each three agent is asked to fill the electronic form to evaluate the five suppliers. The completed forms are collected by TMA and presented in Table 3. Moreover, importance weights of each trust criterion, which is assigned by TMA, are also expressed in this table.

Also the linguistic weights of agents, with respect to their frequency of interaction with suppliers, are collected by the electronic forms (Table 4).

Two phases have been defined for data aggregation. At the first Phase of aggregation, concerning the conversions of the linguistic data of agents' judgments and the frequency of interactions to IT2-FSs, the LWA operator is applied to provide an evaluation of each supplier for each of the defined trust criterion. Figure 6 demonstrates the evaluation of supplier 2 based on four trust criteria.

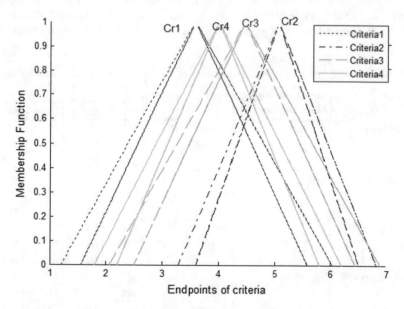

Fig. 6 Statistics endpoints of all criteria for supplier 2

Also at the second phase of this step (Step 2), another LWA operator computes the overall evaluation for each supplier concerning the importance weights of all criteria as IT2-FSs (Fig. 7). The result of the overall aggregation of each supplier is depicted in Fig. 8.

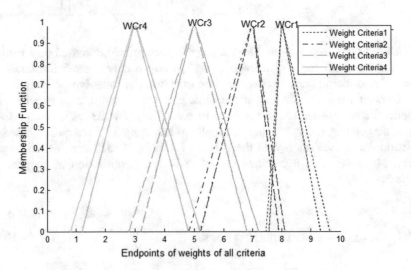

Fig. 7 Statistics endpoints of weights of all criteria for supplier 2

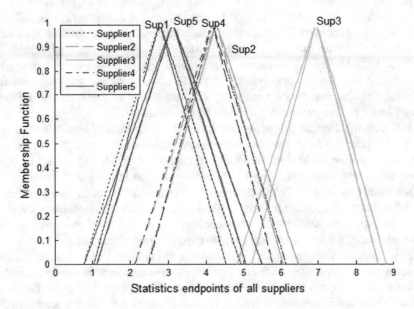

Fig. 8 Statistics endpoints of all suppliers

Table 5 The result of prioritizing suppliers based on trust criteria using type-2 fuzzy sets

Criterion	Scores	Ranks
Supplier 1	2.8103	5
Supplier 2	4.2791	2
Supplier 3	6.9003	1
Supplier 4	4.1517	3
Supplier 5	3.1316	4

Finally in the Step 3 of the Stage 2 suppliers are prioritized based on a method proposed by Asan and his colleagues in [4] using α-cuts Table 5 illustrates the result of prioritizing the suppliers based on trust criteria, using type-2 fuzzy sets. It can be seen from the Table that the most and the least trustable suppliers are supplier 3 and supplier 1 with the value of 6.9003 and 2.8103 respectively. Based on the descending order, supplier 2, supplier 4 and supplier 5 are ranked based on trustworthiness evaluation with the values of 4.2791, 4.1517 and 3.1316 respectively. Moreover, Fig. 8 confirms the result of the prioritization of suppliers in Table 5.

7 Conclusion and Future Work

E-procurement has been recognized as a strategic tool for improving the competitiveness of the firms and generating scale economies for the both sellers and buyers. In an agent based e-procurement system, intelligent agents exhibit good capability to function on behalf of the buyers to look for a most satisfying seller (or supplier). In this context, it is crucial to protect buyers from any possible unsatisfied conditions which are commonly due the existence of uncertain and vague characters in their interrelation. Presence of inter-organizational trust in the buyer-seller relationship was found not only reducing such uncertainty and vagueness characteristics, but also significantly reducing the complexity of the inter-firm relations which could in turn reduce supply risks in this context.

Based on this finding, this paper considered the utilization of inter-organizational trust concept in the agent based e-procurement environment, through proposing an effective supplier evaluation and ranking method. Considering an agent based e-procurement system consisting of specific agents for the required functions, the paper, further included a Trust Management Agent (ATM) which is responsible to establish inter-organizational trust in the buyer-supplier relationship. ATM is to evaluate some pre-qualified candidate suppliers and to rank them on some particular trust criteria. Characteristics inherent in the inter-organizational trust evaluation process, such as: uncertainty and vagueness in the data, use of collective opinions from experts or other agent in the community, were considered in the determination of the solution approach. With regard to the solution approach, Type-2 fuzzy sets proved to be most suitable in dealing trust evaluation process efficiently. In this regard, a new evaluation process based on hierarchical linguistic weighted averaging sets was proposed. The solution method was then illustrated through a simple example which clarifies the suitability as well as the simplicity of the proposed method for the category of the defined problem.

For the future work we will concentrate on changing points in the LWA, which can be estimated by heuristic methods instead of KM or EKM algorithms, also using real data that can help to validate and verify the problem.

References

1. G. Acampora, D. Alghazzawi, H. Hagras, A. Vitiello, An interval type-2 fuzzy logic based framework for reputation management in Peer-to-Peer e-commerce. Inf. Sci. **333**, 88–107 (2016)
2. J.M. Alvarez-Rodríguez, J.E. Labra-Gayo, P.O. de Pablos, New trends on e-Procurement applying semantic technologies: current status and future challenges. Comput. Ind. **65**(5), 800–820 (2014)
3. G. Anders, J.-P. Steghöfer, F. Siefert, W. Reif, Patterns to measure and utilize trust in multi-agent systems, in *Fifth IEEE Conference on Self-Adaptive and Self-Organizing Systems Workshops* (2011)
4. U. Asan, A. Soyer, E. Bozdag, An interval type-2 fuzzy prioritization approach to project risk assessment. Multiple-Valued Logic Soft Comput. **26**(6), 541–577 (2016)
5. O. Castillo, P. Melin, J. Kacprzyk, W. Pedrycz, Type-2 fuzzy logic: theory and applications, in *2007 IEEE International Conference on Granular Computing, GRC 2007* (IEEE, 2007)
6. C.F. Cheung, W.M. Wang, V. Lo, W. Lee, An agent-oriented and knowledge-based system for strategic e-procurement. Expert Syst. **21**(1), 11–21 (2004)
7. P.M. Doney, J.P. Cannon, An examination of the nature of trust in buyer-seller relationships. J. Mark. 35–51 (1997)
8. W. Dong, F. Wong, Fuzzy weighted averages and implementation of the extension principle. Fuzzy Sets Syst. **21**(2), 183–199 (1987)
9. B. Hadikusumo, S. Petchpong, C. Charoenngam, Construction material procurement using internet-based agent system. Autom. Constr. **14**(6), 736–749 (2005)
10. M. He, N.R. Jennings, H.-F. Leung, On agent-mediated electronic commerce. IEEE Trans. Knowl. Data Eng. **15**(4), 985–1003 (2003)
11. H. Huang, G. Zhu, S. Jin, Revisiting trust and reputation in multi-agent systems, in *2008 ISECS International Colloquium on Computing, Communication, Control, and Management, CCCM'08* (IEEE, 2008)
12. C.K. Lee, H.C. Lau, G.T. Ho, W. Ho, Design and development of agent-based procurement system to enhance business intelligence. Expert Syst. Appl. **36**(1), 877–884 (2009)
13. P. Maes, R.H. Guttman, A.G. Moukas, Agents that buy and sell. Commun. ACM **42**(3), 81–ff (1999)
14. J. Mendel, D. Wu, *Perceptual Computing: Aiding People in Making Subjective Judgments* (Wiley, 2010)
15. J.M. Mendel, On KM algorithms for solving type-2 fuzzy set problems. IEEE Trans. Fuzzy Syst. **21**(3), 426–446 (2013)
16. J.M. Mendel, R.B. John, Type-2 fuzzy sets made simple. IEEE Trans. Fuzzy Syst. **10**(2), 117–127 (2002)
17. J.M. Mendel, R.I. John, F. Liu, Interval type-2 fuzzy logic systems made simple. IEEE Trans. Fuzzy Syst. **14**(6), 808–821 (2006)

18. Z. Moattar Husseini, M.F. Zarandi, S. Moattar Husseini, Trust evaluation for buyer-supplier relationship concerning fuzzy approach, in *2015 Annual Conference of the North American Fuzzy Information Processing Society (NAFIPS) Held Jointly with 2015 5th World Conference on Soft Computing (WConSC)* (IEEE, 2015)
19. N.S. Raghavan, M. Prabhu, Object-oriented design of a distributed agent-based framework for e-Procurement. Prod. Plann. Control **15**(7), 731–741 (2004)
20. S.D. Ramchurn, D. Huynh, N.R. Jennings, Trust in multi-agent systems. Knowl. Eng. Rev. **19** (01), 1–25 (2004)
21. D. Rosaci, G.M. Sarnè, Multi-agent technology and ontologies to support personalization in B2C E-Commerce. Electron. Commer. Res. Appl. **13**(1), 13–23 (2014)
22. S. Ruohomaa, L. Kutvonen, Trust management survey, in *Trust Management* (Springer, 2005), pp. 77–92
23. J. Sabater, C. Sierra, Reputation and social network analysis in multi-agent systems, in *Proceedings of the First International Joint Conference on Autonomous Agents and Multiagent Systems: Part 1* (ACM, 2002)
24. S. Schmidt, R. Steele, T.S. Dillon, E. Chang, Fuzzy trust evaluation and credibility development in multi-agent systems. Appl. Soft Comput. **7**(2), 492–505 (2007)
25. R. Seppänen, K. Blomqvist, S. Sundqvist, Measuring inter-organizational trust—a critical review of the empirical research in 1990–2003. Ind. Mark. Manage. **36**(2), 249–265 (2007)
26. S.X. Sun, J. Zhao, H. Wang, An agent based approach for exception handling in e-procurement management. Expert Syst. Appl. **39**(1), 1174–1182 (2012)
27. M. Witkowski, A. Artikis, J. Pitt, Experiments in building experiential trust in a society of objective-trust based agents, in *Trust in Cyber-societies* (Springer, 2001), pp. 111–132
28. D. Wu, J.M. Mendel, Aggregation using the linguistic weighted average and interval type-2 fuzzy sets. IEEE Trans. Fuzzy Syst. **15**(6), 1145–1161 (2007)
29. D. Wu, J.M. Mendel, Corrections to "Aggregation using the linguistic weighted average and interval type-2 fuzzy sets". IEEE Trans. Fuzzy Syst. **16**(6), 1664–1666 (2008)
30. Y. Ye, J. Liu, A. Moukas, *Agents in Electronic Commerce* (Springer, 2001)
31. L.A. Zadeh, The concept of a linguistic variable and its application to approximate reasoning —I. Inf. Sci. **8**(3), 199–249 (1975)
32. A. Zaheer, J.D. Harris, Interorganizational trust, in *Handbook of Strategic Alliances*, eds. by O. Shenkar, Jeffrey J. Reuer (2005), pp. 169–197
33. A. Zaheer, B. McEvily, V. Perrone, Does trust matter? Exploring the effects of interorganizational and interpersonal trust on performance. Organ. Sci. **9**(2), 141–159 (1998)
34. Z. Zhang, S. Zhang, A novel approach to multi attribute group decision making based on trapezoidal interval type-2 fuzzy soft sets. Appl. Math. Model. **37**(7), 4948–4971 (2013)
35. P. Zhou, X. Gu, J. Zhang, M. Fei, A priori trust inference with context-aware stereotypical deep learning. Knowl. Based Syst. **88**, 97–106 (2015)

,

Printed in the United States
By Bookmasters